A-LEVEL YEAR 2

STUDENT GUIDE

EDEXCEL

Biology B

Topics 5–7

Energy for biological processes
Microbiology and pathogens
Modern genetics

Mary Jones

PHILIP ALLAN FOR
HODDER
EDUCATION
AN HACHETTE UK COMPANY

Philip Allan, an imprint of Hodder Education, an Hachette UK company, Blenheim Court, George Street, Banbury, Oxfordshire OX16 5BH

Orders

Bookpoint Ltd, 130 Park Drive, Milton Park, Abingdon, Oxfordshire OX14 4SE

tel: 01235 827827

fax: 01235 400401

e-mail: education@bookpoint.co.uk

Lines are open 9.00 a.m.–5.00 p.m., Monday to Saturday, with a 24-hour message answering service. You can also order through the Hodder Education website: www.hoddereducation.co.uk

© Mary Jones 2016

ISBN 978-1-4718-5848-2

First printed 2016

Impression number 5 4 3 2 1

Year 2020 2019 2018 2017 2016

This guide has been written specifically to support students preparing for the Edexcel A-level Biology B examinations. The content has been neither approved nor endorsed by Edexcel and remains the sole responsibility of the author.

Cover photo: Elena Pankova/Fotolia

Typeset by Greenhill Wood Studios

Printed in Italy

Hachette UK's policy is to use papers that are natural, renewable and recyclable products and made from wood grown in sustainable forests. The logging and manufacturing processes are expected to conform to the environmental regulations of the country of origin.

Contents

Content Guidance

Aerobic respiration • Glycolysis • Link reaction and the Krebs cycle • Oxidative phosphorylation • Anaerobic respiration • *Core practical 9: Measuring the rate of respiration* • Photosynthetic pigments • *Core practical 10: Investigating the effects of different wavelengths of light on the rate of photosynthesis* • *Core practical 11: Separating chloroplast pigments by chromatography* • Photosynthesis

Microbial techniques • *Core practical 12: Investigating the rate of growth of bacteria in liquid culture* • *Core practical 13: Isolating individual species from a mixed culture of bacteria using streak plating* • Bacteria as pathogens • Action of antibiotics • Antibiotic resistance • Other pathogenic agents • Problems of controlling endemic diseases • Response to infection

Using gene sequencing • Factors affecting gene expression • Stem cells • Gene technology

Questions & Answers

■ Getting the most from this book

Exam-style questions

Commentary on the questions

Tips on what you need to do to gain full marks, indicated by the icon ⓔ

Sample student answers

Practise the questions, then look at the student answers that follow.

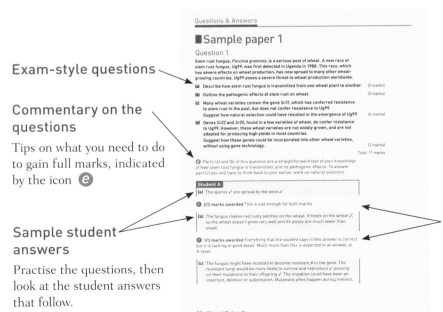

Commentary on sample student answers

Find out how many marks each answer would be awarded in the exam and then read the comments (preceded by the icon ⓔ) following each student answer showing exactly how and where marks are gained or lost.

◼About this book

This book is the third in a series of four covering the Edexcel A-level Biology B specification. It covers Topics 5, 6 and 7:

- Energy for biological processes
- Microbiology and pathogens
- Modern genetics

This guide has two main sections:

- The **Content Guidance** provides a summary of the facts and concepts that you need to know for these three topics.
- The **Questions & Answers** section contains two specimen papers for you to try, one worth 90 marks, similar to paper 1, and one worth 120 marks, similar to paper 3. There are also two sets of answers for each question, one from a student who is likely to get a C grade and another from a student who is likely to get an A grade.

The specification

It is a good idea to have your own copy of the Edexcel Biology B specification. It is you who is going to take this examination, not your teacher, and so it is your responsibility to make sure you know as much about the exam as possible. You can download a copy free from www.edexcel.com.

The A-level examination is made up of three papers:

- **Paper 1** Advanced Biochemistry, Microbiology and Genetics (1 hour 45 minutes, 90 marks)
- **Paper 2** Advanced Physiology, Evolution and Ecology (1 hour 45 minutes, 90 marks)
- **Paper 3** General and Practical Principles in Biology (2 hours 30 minutes, 120 marks)

This book covers content that will be examined in papers 1 and 3 of the A-level examination. Note that all three of the A-level papers also test your knowledge and understanding of Topics 1, 2, 3 and 4. These are covered in the first two student guides in this series.

What is assessed?

It is easy to forget that your examination is not just testing what you *know* about biology — it is also testing your *skills*. It is difficult to overemphasise how important these are.

The Edexcel examination tests three different assessment objectives (AOs). The following table gives a breakdown of the proportion of marks awarded to each assessment objective in the A-level examinations.

Assessment objective	Outline of what is tested	Percentage of marks (A-level)
AO1	Demonstrate knowledge and understanding of scientific ideas, processes, techniques and procedures	31–33
AO2	Apply knowledge and understanding of scientific ideas, processes, techniques and procedures: ■ in a theoretical context ■ in a practical context ■ when handling qualitative data ■ when handling quantitative data	41–43
AO3	Analyse, interpret and evaluate scientific information, ideas and evidence, including in relation to issues, to: ■ make judgements and reach conclusions ■ develop and refine practical design and procedures	25–27

AO1 is about remembering and understanding all the biological facts and concepts you have covered. AO2 is about being able to *use* these facts and concepts in new situations. The examination paper will include questions that contain unfamiliar contexts or sets of data, which you will need to interpret in the light of the biological knowledge you have. When you are revising, it is important that you try to develop your ability to do this, as well as just learning the facts.

AO3 is about practical and experimental biology. A science subject such as biology is not just a body of knowledge. Our knowledge and understanding of biology continues to develop, as scientists find out new information through their research. Sometimes new research means that we have to change our ideas.

You need to develop your skills at doing experiments to test hypotheses, and analysing the results to determine whether the hypothesis is supported or disproved. You need to appreciate why science does not always give us clear answers to the questions we ask. Finally, you will be asked to make judgements and reach conclusions, and need to be able to design and improve experiments and procedures, to produce results we can trust.

Scientific language

Throughout your biology course, and especially in your examination, it is important to use clear and correct biological language. Scientists take great care to use language precisely. If doctors or researchers do not use exactly the correct word when communicating with someone, then what they are saying could easily be misinterpreted.

Biology has a huge number of specialist terms and it is important that you learn them and use them. Your everyday conversational language, or what you read in the newspaper or hear on the radio, is often not the kind of language required in a biology exam. Be precise and careful in what you write, so that an examiner cannot possibly misunderstand you.

The examination

Time

In all of the examinations, the mark allocation works out at around 1 minute per mark. When you are trying out a test question, time yourself. Are you working too fast? Or are you taking too long? Get used to what it feels like to work at around a mark-a-minute rate.

It is not a bad idea to spend one of those minutes just skimming through the exam paper before you start writing. Maybe one of the questions looks as though it is going to need a bit more of your time than the others. If so, make sure you leave a little bit of extra time for it.

Read the question carefully

This sounds obvious, but students lose large numbers of marks by not doing it.

- There is often vital information at the start of the question that you will need in order to answer the questions themselves. Do not just jump straight to the first set of answer lines and start writing — start reading at the beginning. Examiners are usually careful not to give you unnecessary information, so if it is there it is probably needed. You may like to use a highlighter to pick out any particularly important bits of information in the question.
- Look carefully at the command words (the ones right at the start of the question) and do what they say. For example, if you are asked to *explain* something, you will not get many marks — perhaps none at all — if you *describe* it instead. You can find all these words in an appendix near the end of the specification document.

Depth and length of answer

The examiners will give you two useful guidelines about how much you need to write:

- **The number of marks**. Obviously, the more marks allocated, the more information you need to give. If there are 2 marks, then you will need to give two different pieces of information in order to get both of them. If there are 5 marks, you will need to write much more.
- **The number of lines**. This is not such a useful guideline as the number of marks, but it can still help you to know how much to write. If you find your answer won't fit on the lines, then you probably haven't focused sharply enough on the question. The best answers are short and precise.

Mathematical skills

Like all of the sciences, biology uses mathematics extensively. The specification contains an appendix that lists and describes the mathematical techniques that you will need to be familiar with. You will probably have met most of these before, but make sure that you are confident with all of them. If there are any of which you are uncertain, then do your best to improve your skills in them early on in the course — do not leave it until the last minute, just before the exam. The more you practise your maths skills, the more relaxed you will be about them in the exam.

Mathematical skills can be tested in any of the A-level papers. However, they are particularly likely to be tested in paper 3.

Content Guidance

■ Topic 5 Energy for biological processes

Aerobic respiration

ATP

ATP stands for adenosine triphosphate (Figure 1). ATP is a phosphorylated nucleotide — it has a similar structure to the nucleotides that make up RNA. However, it has three phosphate groups attached to it instead of one.

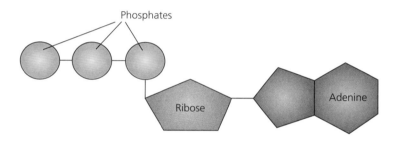

Figure 1 An ATP molecule

ATP is used as the energy currency in every living cell. ATP molecules contain a lot of chemical energy. When an ATP molecule is hydrolysed, losing one of its phosphate groups, some of this energy is released and can be used by the cell. In this process, the ATP is converted to ADP (adenosine diphosphate). ATP is used as a source of energy for metabolic reactions.

All cells make ATP by **respiration**. Cellular respiration is a series of enzyme-controlled reactions in which energy in organic substances, often glucose, is released in a series of steps and transferred to ATP. Some energy is also released as heat.

The stages of aerobic respiration

In aerobic respiration, oxygen is involved. Glucose, $C_6H_{12}O_6$, (or another respiratory substrate) is split to release carbon dioxide as a waste product. The hydrogen from the glucose is combined with atmospheric oxygen. This releases a large amount of energy, which is used to drive the synthesis of ATP.

Aerobic respiration takes place in four stages:
- **glycolysis**, in which glucose is oxidised to pyruvate
- **the link reaction**, in which pyruvate is converted to acetyl coenzyme A

- **the Krebs cycle**, in which the 2-carbon acetyl coenzyme A enters a cyclical series of reactions
- **oxidative phosphorylation**, in which hydrogens released during the previous stages are used to convert ADP to ATP

Glycolysis takes place in the cytoplasm of a cell, but the remaining three stages all take place inside mitochondria.

Summary

After studying this topic, you should be able to:
- state that cellular respiration yields ATP and heat
- state that ATP is used as a source of energy for metabolic reactions
- list the different stages of aerobic respiration

Glycolysis

Glycolysis (Figure 2) is the first stage of respiration. It takes place in the cytoplasm.

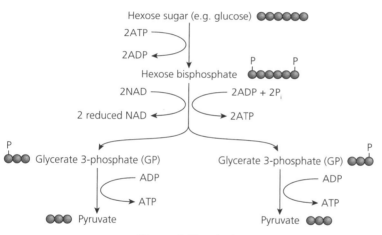

Figure 2 Glycolysis

A glucose (or other monosaccharide) molecule is phosphorylated, as two ATPs donate phosphate to it.

This produces a hexose bisphosphate molecule, which splits into two molecules of glycerate 3-phosphate, GP. This involves the removal of hydrogens, which are taken up by a coenzyme called **NAD**. This produces reduced NAD, sometimes written as NADH.

Each GP molecule is then converted to a **pyruvate** molecule. (Pyruvate may also be referred to as pyruvic acid.) During this step, the phosphate groups from the GP are added to ADP to make ATP.

Overall, two molecules of ATP are used and four are made during glycolysis of one glucose molecule, making a net gain of two ATPs per glucose molecule.

Knowledge check 2

Suggest why glucose must be phosphorylated before it is split.

Summary

After studying this topic, you should be able to:
- describe how and why glucose is phosphorylated
- state that phosphorylated glucose is broken down to glycerate 3-phosphate (GP)
- explain how reduced coenzyme (NADH) and ATP are produced as the GP is converted to pyruvate

Link reaction and the Krebs cycle

The link reaction

If oxygen is available, the pyruvate now moves into the matrix of a mitochondrion (Figure 3), where the link reaction (Figure 4) and the Krebs cycle take place. During these processes, the glucose is completely oxidised.

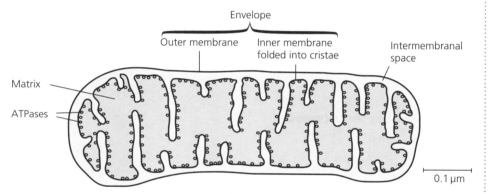

Figure 3 A mitochondrion

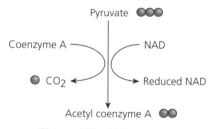

Figure 4 The link reaction

Carbon dioxide is removed from the pyruvate. This carbon dioxide diffuses out of the mitochondrion and out of the cell. Hydrogen is also removed from the pyruvate, and is picked up by NAD, producing reduced NAD. This converts pyruvate into a two-carbon compound. This immediately combines with **coenzyme A** to produce acetyl coenzyme A.

The Krebs cycle

Acetyl coenzyme A has two carbon atoms. It combines with a four-carbon compound to produce a six-carbon compound. This is gradually converted to the four-carbon

Knowledge check 3

From which substance do the carbon atoms in CO_2 in expired air originate? Use the flow diagrams in Figures 2 and 4 to work this out.

compound again through a series of enzyme-controlled steps. These steps all take place in the matrix of the mitochondrion, and each is controlled by specific enzymes.

During this process, more carbon dioxide is released and diffuses out of the mitochondrion and out of the cell. More hydrogens are released and picked up by NAD and another coenzyme called FAD. This produces reduced NAD (NADH) and reduced FAD (FADH). At one point, ATP is produced (Figure 5).

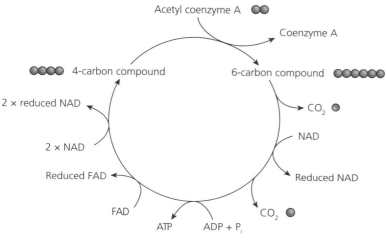

Figure 5 The Krebs cycle

Summary

After studying this topic, you should be able to:
- state that the link reaction and Krebs cycle take place in the matrix of a mitochondrion
- describe how the link reaction and Krebs cycle result in the complete oxidation of pyruvate
- state that carbon dioxide, reduced coenzyme (NADH) and ATP are produced in these reactions

Oxidative phosphorylation

The hydrogens picked up by NAD and FAD are now split into electrons and protons. The electrons are passed along a series of molecules in the inner membrane of the mitochondrion, called the **electron transport chain**.

As the electrons move along the chain, they lose energy. This energy is used to actively transport hydrogen ions from the matrix of the mitochondrion, across the inner membrane and into the space between the inner and outer membranes. This builds up a high concentration of hydrogen ions in this space.

The hydrogen ions are then allowed to diffuse back into the matrix through special channel proteins that work as **ATP synthase** enzymes. The movement of the hydrogen ions down their concentration gradient and electrical gradient, through the ATP synthases, provides enough energy to cause ADP and inorganic phosphate to combine to make ATP (Figure 6).

The active transport and subsequent diffusion of the hydrogen ions across the inner mitochondrial membrane is known as **chemiosmosis**.

Exam tip

Do not confuse chemiosmosis with osmosis. Osmosis is a passive process involving the diffusion of water molecules through a partially permeable membrane.

The electron transport chain provides energy to transport hydrogen ions from the matrix to the space between the inner and outer membranes.

The hydrogen ions diffuse back through ATP synthase, providing energy to make ATP.

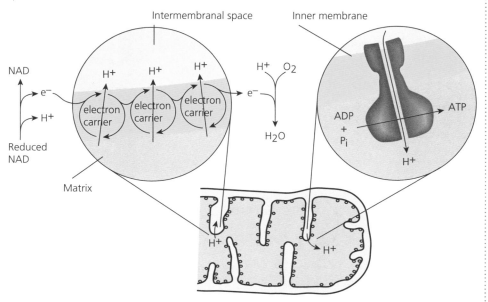

Figure 6 Oxidative phosphorylation

At the end of the chain, the electrons reunite with protons. They combine with oxygen, to produce water. This is why oxygen is required in aerobic respiration — it acts as the final acceptor for the hydrogens removed from the respiratory substrate during glycolysis, the link reaction and the Krebs cycle.

Knowledge check 4

Suggest why the process described in this section is known as oxidative phosphorylation.

Summary

After studying this topic, you should be able to:
- state that the electron transport chain is found in the inner mitochondrial membrane
- describe the role of the electron transport chain and ATP synthase in generating ATP
- explain the importance of oxygen as a final electron acceptor, forming water
- explain what is meant by the term chemiosmosis

Anaerobic respiration

If oxygen is not available, oxidative phosphorylation cannot take place, as there is nothing to accept the electrons and protons at the end of the electron transport chain. This means that reduced NAD is not reoxidised, so the mitochondrion quickly runs out of the NAD or FAD that can accept hydrogens from the Krebs cycle reactions. So the Krebs cycle and the link reaction come to a halt.

Glycolysis, however, can still continue, so long as the pyruvate produced at the end of it can be removed and the reduced NAD can be converted back to NAD.

The lactate pathway

In animals, NAD is regenerated by converting the pyruvate to **lactate** (Figure 7).

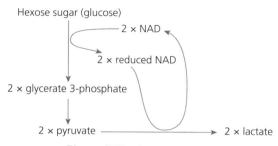

Figure 7 The lactate pathway

The lactate that is produced (usually in muscles) diffuses into the blood and is carried in solution in the blood plasma to the liver. Here, liver cells convert it back to pyruvate. This requires oxygen, so extra oxygen is required after exercise has finished. The extra oxygen is known as the oxygen debt. Later, when the exercise has finished and oxygen is available again, some of the pyruvate in the liver cells is oxidised through the link reaction, the Krebs cycle and the electron transport chain. Some of the pyruvate is reconverted to glucose in the liver cells, and this can be released into the blood or converted to glycogen and stored.

When oxygen is in short supply, and muscles are using the lactate pathway, a build-up of lactate in the muscle tissue may contribute to muscle fatigue.

The ethanol pathway

In plants and yeast, the pyruvate is converted to **ethanol** (Figure 8).

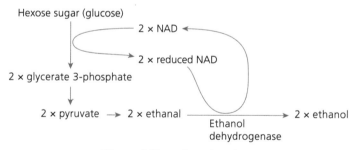

Figure 8 The ethanol pathway

ATP yield in aerobic and anaerobic respiration

Only small amounts of ATP are produced when one glucose molecule undergoes anaerobic respiration. This is because only glycolysis is completed. The Krebs cycle and oxidative phosphorylation, which produce most ATP, do not take place.

The precise number of molecules of ATP produced in aerobic respiration of one glucose molecule varies between different organisms and different cells, but is usually between 30 and 32 molecules. Figure 9 summarises the four stages in aerobic respiration, and also shows the ATP yields of each stage.

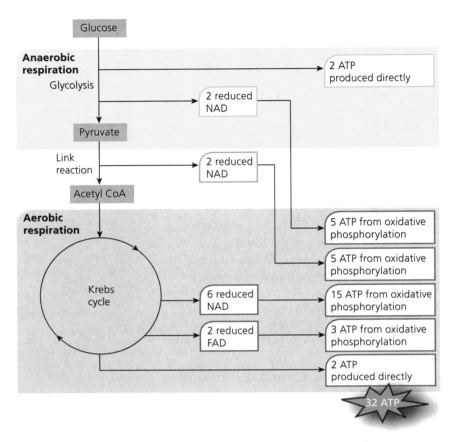

Figure 9 A summary of the four stages of respiration

Core practical 9

Measuring the rate of respiration

You can measure the rate of uptake of oxygen using a **respirometer** (Figure 10).

The organisms to be investigated are placed in one test tube, and non-living material of the same mass is placed in the other test tube. Soda lime is placed in each test tube, to absorb all carbon dioxide. Cotton wool prevents contact of the soda lime with the organisms.

Coloured fluid is poured into the reservoir of each manometer and allowed to flow into the capillary tube. It is essential that there are no air bubbles. You must end up with exactly the same quantity of fluid in the two manometers.

Two rubber bungs are now taken, fitted with tubes, as shown in Figure 10. Close the spring clips. Attach the manometers to the bent glass tubing, ensuring a totally airtight connection. Next, place the bungs into the tops of the test tubes.

Open the spring clips. (This allows the pressure throughout the apparatus to equilibrate with atmospheric pressure.) Note the level of the manometer fluid in each tube. Close the clips. Each minute, record the level of the fluid in each tube.

continued

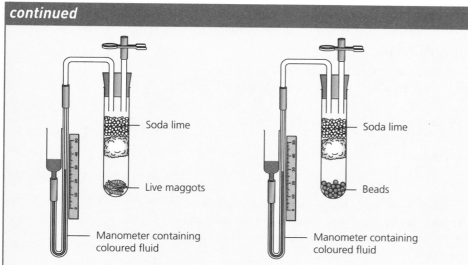

Figure 10 A respirometer

As the organisms respire aerobically, they take oxygen from the air around them and give out carbon dioxide. The removal of oxygen from the air inside the tube reduces the volume and pressure, causing the manometer fluid to move towards the organisms. The carbon dioxide given out is absorbed by the soda lime.

You would not expect the manometer fluid in the tube with no organisms to move, but it may do so because of temperature changes. This allows you to control for this variable, by subtracting the distance moved by the fluid in the control manometer from the distance moved in the experimental manometer (connected to the living organisms), to give you an adjusted distance moved.

Calculate the mean (adjusted) distance moved by the manometer fluid per minute. If you know the diameter of the capillary tube, you can convert the distance moved to a volume:

volume of liquid in a tube = length × πr^2

This gives you a value for the volume of oxygen absorbed by the organisms per minute.

You can compare rates of respiration at different temperatures by standing the apparatus in a water bath. You can also compare rates of respiration using different concentrations of substrate by using a suspension of yeast in glucose solution, varying the concentration of glucose.

Knowledge check 5

Predict and explain what would happen to the levels of fluid in the manometers if no soda lime was used.

Knowledge check 6

Predict and explain what would happen to the levels of fluid in the manometers if the organisms were respiring anaerobically.

Summary

After studying this topic, you should be able to:
- explain that anaerobic respiration is the partial breakdown of glucose (or other hexoses) in the absence of oxygen, producing a small yield of ATP
- compare the ATP yields from one molecule of hexose in aerobic and anaerobic conditions
- describe the lactate pathway in animals, and how it can affect muscle contraction
- describe the ethanol pathway in plants
- explain how the lactate pathway and ethanol pathway enable glycolysis to continue
- use a respirometer to investigate factors affecting the rate of aerobic or anaerobic respiration

Photosynthetic pigments

An overview of photosynthesis

Photosynthesis is a series of reactions in which energy transferred as light is transformed to chemical energy. Energy from light is trapped by chlorophyll, and this energy is then used to:

- split apart the strong bonds in water molecules to release hydrogen
- produce ATP
- reduce a substance called NADP

NADP, like NAD, is a coenzyme.

The ATP and reduced NADP are then used to add hydrogen to carbon dioxide, to produce carbohydrate molecules such as glucose. These complex organic molecules contain some of the energy that was originally in the light. The oxygen from the split water molecules is a waste product, and is released into the air.

There are many different steps in photosynthesis, which can be divided into two main stages — the light-dependent stage and the light-independent stage.

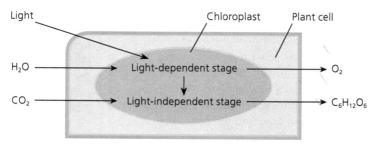

Figure 11 An overview of photosynthesis

Exam tip

Students often say that carbon dioxide is changed to oxygen in photosynthesis. Look at the diagram and see for yourself why this is not correct.

Chloroplast pigments

A **pigment** is a substance that absorbs light of some wavelengths but not others. The wavelengths that is does not absorb are reflected from it.

Chlorophyll is the main pigment contained in chloroplasts. It looks green because it reflects green light. Other wavelengths (colours) of light are absorbed.

Figure 12 shows the wavelengths of light that are absorbed by the various pigments found in chloroplasts. These graphs are called **absorption spectra**.

If we shine light of various wavelengths on chloroplasts, we can measure the rate at which they give off oxygen. This graph is called an **action spectrum** (Figure 13).

Chlorophyll a is the most abundant pigment in most plants. Its absorption peaks are 430 nm (blue) and 662 nm (red). It emits an electron when it absorbs light.

Chlorophyll b is similar to chlorophyll a, but its absorption peaks are 453 nm and 642 nm. It has a similar role to chlorophyll a, but is not as abundant.

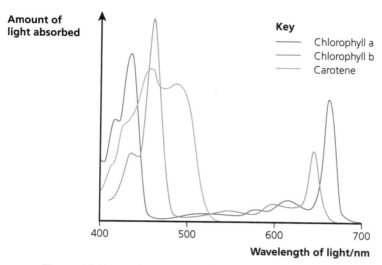

Figure 12 Absorption spectra for chloroplast pigments

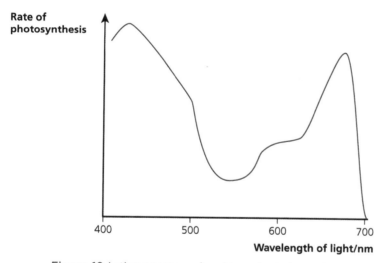

Figure 13 Action spectrum for chloroplast pigments

Carotenoids are accessory pigments. They are orange pigments that protect chlorophyll from damage by the formation of single oxygen atoms (free radicals). They can also absorb wavelengths of light that chlorophyll cannot absorb, and pass on some of the energy from the light to chlorophyll.

Xanthophylls are also accessory pigments, capturing energy from wavelengths of light that are not absorbed by chlorophyll.

Knowledge check 7

Explain the similarities between an absorption spectrum and an action spectrum.

Core practical 10

Investigating the effects of different wavelengths of light on the rate of photosynthesis

One way to measure the rate of photosynthesis is to measure the rate at which oxygen is given off by an aquatic plant such as *Elodea* or *Cabomba*. There are various ways in which oxygen can be collected and measured. One method is shown in Figure 14.

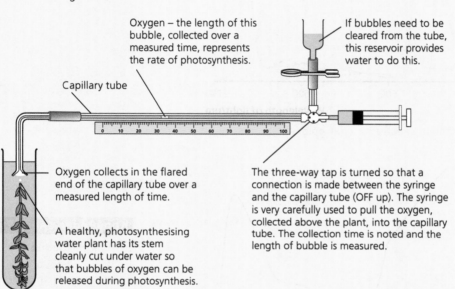

Oxygen – the length of this bubble, collected over a measured time, represents the rate of photosynthesis.

If bubbles need to be cleared from the tube, this reservoir provides water to do this.

Capillary tube

Oxygen collects in the flared end of the capillary tube over a measured length of time.

A healthy, photosynthesising water plant has its stem cleanly cut under water so that bubbles of oxygen can be released during photosynthesis.

The three-way tap is turned so that a connection is made between the syringe and the capillary tube (OFF up). The syringe is very carefully used to pull the oxygen, collected above the plant, into the capillary tube. The collection time is noted and the length of bubble is measured.

Figure 14 Apparatus for measuring the rate of photosynthesis

Alternatively, you can make calcium alginate balls containing green algae and place them in hydrogencarbonate indicator solution. As the algae photosynthesise, they take in carbon dioxide, which causes the pH around them to increase. The indicator changes from orange, through red to magenta. You can find details of this technique at www.tinyurl.com/pfocp6u.

A completely different method of measuring the rate of photosynthesis is to use a redox indicator. A redox indicator is a substance that changes colour when it is oxidised or reduced. Photosynthesis, like respiration, involves the acceptance of hydrogen by a coenzyme. This occurs during the light-dependent stage (page 21), when hydrogen is accepted by NADP. We can investigate the rate at which this occurs by adding a redox indicator, such as DCPIP, to a suspension of chloroplasts. The indicator takes up the hydrogen ions that are produced as the light-dependent stage occurs in the chloroplasts, and loses its colour. This is called the Hill reaction. The rate at which the colour is lost is determined by the rate of the light-dependent stage.

You can find a detailed protocol for carrying out this investigation at www.tinyurl.com/ojnqqg3.

continued

Whichever technique is used, you should change one factor (your independent variable) while keeping all others constant (the control variables). The independent variable is the wavelength of light. You can vary this by placing coloured filters between the light source and the plant, algae or chloroplast extract. Each filter will allow only light of certain wavelengths to pass through.

The dependent variable will be the rate at which oxygen is given off (measured by the volume of oxygen collected per minute in the capillary tube) or the rate at which carbon dioxide is used (measured by the rate of change of colour of the hydrogencarbonate indicator solution).

Other variables that can affect the rate of photosynthesis, such as temperature or light intensity, should be kept constant.

> **Knowledge check 8**
>
> How can you keep temperature constant during this investigation?

Core practical 11

Separating chloroplast pigments by chromatography

Chromatography is a method of separation that relies on the different solubilities of different solutes in a solvent. A mixture of chlorophyll pigments is dissolved in a solvent, and then a small spot is placed onto chromatography paper. The solvent gradually moves up the paper, carrying the solutes with it. The more soluble the solvent, the further up the paper it is carried.

There are various methods. The one described here uses thin layer chromatography on specially prepared strips instead of paper. Only an outline of the procedure is given in Figure 15, so you cannot use these instructions to actually carry out the experiment. You can find more details about this technique at www-saps.plantsci.cam.ac.uk/worksheets/ssheets/ssheet10.htm.

Cut a TLC plate into narrow strips, about 1.25 cm wide, so they fit into a test tube. Do not put your fingers on the powdery surface.

Put 2 or 3 grass leaves on a slide. Using another slide scrape the leaves to extract cell contents.

Add 6 drops of propanone to the extract and mix.

Transfer the mixture to a watch glass. Allow this to dry out almost completely — a warm air flow will speed this up.

Transfer tiny amounts of the concentrated extract onto a spot 1 cm from one end of the TLC strip.

Touch very briefly with the fine tip of the brush and let that spot dry before adding more. Keep the spot to 1 mm diameter if you can. The final spot, called the origin, should be small but dark green.

Figure 15 Chromatography

continued

Measure the distance from the start line to the solvent front. Measure the distances of each pigment spot from the start line. For each spot, calculate the *Rf* value:

$$Rf = \frac{\text{distance from start line to pigment spot}}{\text{distance from start line to solvent front}}$$

You can use the *Rf* values to help you to identify the pigments. *Rf* values differ depending on the solvent you have used, but typical values might be:

- chlorophyll a   0.60
- chlorophyll b   0.50
- carotene   0.95
- xanthophyll   0.35

You may also see a small grey spot with an *Rf* value of about 0.8. This is phaeophytin, which is not really a chloroplast pigment but is a breakdown product generated during the extraction process.

Summary

After studying this topic, you should be able to:
- explain what is meant by absorption spectra and action spectra
- explain why many plants have a variety of photosynthetic pigments
- investigate how the wavelength of light affects the rate of photosynthesis
- use chromatography to investigate the different chloroplast pigments present in leave

Photosynthesis

Chloroplasts

Photosynthesis takes place inside chloroplasts (Figure 16). Like mitochondria, these are organelles surrounded by two membranes, called an envelope. They are found in mesophyll cells in leaves. Palisade mesophyll cells contain most chloroplasts, but chloroplasts are also found in spongy mesophyll cells. Guard cells also contain chloroplasts.

The membranes inside a chloroplast are called **lamellae**, and it is here that the light-dependent stages take place. The membranes contain chlorophyll molecules, arranged in groups called **photosystems**. There are two kinds of photosystem, PSI and PSII, each of which contains slightly different kinds of chlorophyll.

There are enclosed spaces between pairs of membranes, forming fluid-filled sacs called **thylakoids**. These are involved in **photophosphorylation** — the formation of ATP using energy from light. Thylakoids are often arranged in stacks called **grana** (singular: granum).

The 'background material' of the chloroplast is called the **stroma**, and this is where the light-independent stage takes place.

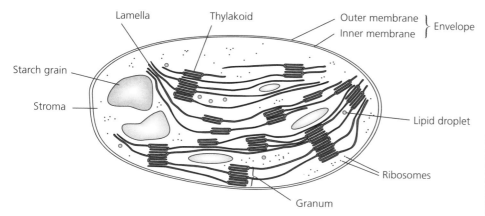

Figure 16 Structure of a chloroplast

Chloroplasts often contain starch grains and lipid droplets. These are stores of energy-containing substances that have been made in the chloroplast but are not immediately needed by the cell or by other parts of the plant.

The light-dependent stage

Chlorophyll molecules in PSI and PSII absorb light energy. The energy excites electrons, raising their energy level so that they leave the chlorophyll.

PSII contains an enzyme that splits water when activated by light. This reaction is called **photolysis** ('splitting by light'). The water molecules are split into oxygen and hydrogen atoms. Each hydrogen atom then loses its electron, to become a positively charged hydrogen ion (proton), H^+. The electrons are picked up by the chlorophyll in PSII, to replace the electrons it lost. The oxygen atoms join together to form oxygen molecules, which diffuse out of the chloroplast and into the air around the leaf.

$$2H_2O \xrightarrow{\text{light}} 4H^+ + 4e^- + O_2$$

The electrons emitted from PSII are picked up by electron carriers in the membranes of the thylakoids. They are passed along a chain of these carriers, losing energy as they go. In a similar way to oxidative phosphorylation in mitochondria, the energy they lose is used to make ADP combine with a phosphate group, producing ATP. This is called **photophosphorylation**. At the end of the electron carrier chain, the electron is picked up by PSI. This replaces the electron the chlorophyll in PSI had lost when it absorbed the light energy.

The electrons from PSI are passed along a different chain of carriers to NADP. The NADP also picks up the hydrogen ions from the split water molecules. The NADP becomes reduced NADP.

We can show all of this in a diagram called the Z-scheme (Figure 17). The higher up the diagram, the higher the energy level. If you follow one electron from a water molecule, you can see how it:

- is taken up by PSII
- has its energy raised as the chlorophyll in PSII absorbs light energy
- loses some of this energy as it passes along the electron carrier chain

Knowledge check 10

Compare the structures of a mitochondrion and a chloroplast.

Exam tip

The supply of energy for the light-dependent stage comes from light. These reactions, therefore, are not affected by temperature, unlike most other reactions in living organisms.

- is taken up by PSI
- has its energy raised again as the chlorophyll in PSI absorbs light energy
- becomes part of a reduced NADP molecule

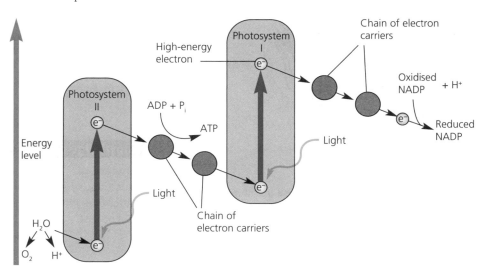

Figure 17 Summary of the light-dependent stage of photosynthesis — the Z-scheme

At the end of this process, two new high-energy substances have been made. These are ATP and reduced NADP. Both of them will now be used in the next stage of photosynthesis, the light-independent stage.

Cyclic and non-cyclic photophosphorylation

The sequence of events just described and shown in Figure 17 is known as **non-cyclic photophosphorylation**.

There is an alternative pathway for the electron that is emitted from PSI. It can simply be passed along the electron transport chain, then back to PSI again. ATP is produced as it moves along the electron transport chain (photophosphorylation). However, no reduced NADP is produced. This is called **cyclic photophosphorylation**.

The light-independent stage

This takes place in the stroma of the chloroplast, where the enzyme ribulose bisphosphate carboxylase, usually known as **RUBISCO**, is found.

Carbon dioxide diffuses into the stroma from the air spaces within the leaf. It enters the active site of RUBISCO, which combines it with a 5-carbon compound called ribulose bisphosphate, **RuBP**. The products of this reaction are two 3-carbon molecules, glycerate 3-phosphate, **GP**. The combination of carbon dioxide with RuBP is called **carbon fixation**.

Energy from ATP and hydrogen from reduced NADP are then used to convert the GP into triose phosphate, TP. This is sometimes known as glyceraldehyde 3-phosphate, or **GALP** for short. This is the first carbohydrate produced in photosynthesis.

Most of the GALP is used to produce ribulose bisphosphate, so that more carbon dioxide can be fixed. The rest is used to make monosaccharides such as glucose, or whatever other organic substances the plant cell requires. These include polysaccharides such as starch for energy storage and cellulose for making cell walls, sucrose for transport, amino acids for making proteins, lipids for energy storage and nucleotides for making DNA and RNA.

This cyclical series of reactions is known as the **Calvin cycle** (Figure 18).

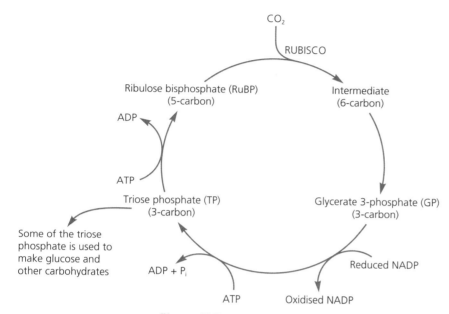

Figure 18 The Calvin cycle

Exam tip

Some old textbooks refer to the light-independent reactions as the 'dark reactions'. Do not use this term. The reactions can happen perfectly well in light — they just don't need light.

Knowledge check 13

Where do the reduced NADP and the ATP, used in the Calvin cycle, come from?

Limiting factors in photosynthesis

The rate at which photosynthesis takes place is directly affected by several environmental factors.

- **Light intensity** This affects the rate of the light-dependent reaction, which is driven by energy transferred in light rays.
- **Temperature** This affects the rate of the light-independent reaction. At higher temperatures, molecules have more kinetic energy so collide more often and are more likely to react when they do collide. (The rate of the light-dependent reaction is not affected by temperature, as the energy that drives this process is light energy, not heat energy.)
- **Carbon dioxide concentration in the atmosphere** Carbon dioxide is a reactant in photosynthesis. Normal air contains only about 0.04% carbon dioxide.

If the level of any one of these factors is too low, then the rate of photosynthesis will be reduced. The factor that has the greatest effect in reducing the rate is said to be the **limiting factor** (Figure 19).

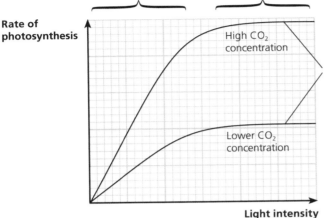

Over this range of the graphs, light intensity is the limiting factor. If light intensity increases, then the rate of photosynthesis increases.

Over this range of the graph, light intensity is not a limiting factor. If light intensity increases, there is no effect on the rate of photosynthesis. Some other factor is limiting the rate.

Rate of photosynthesis

High CO_2 concentration

Here, carbon dioxide is the limiting factor. We can tell this is so because when the concentration of carbon dioxide is increased, the rate of photosynthesis increases (see top curve).

Lower CO_2 concentration

Light intensity

Figure 19 Limiting factors for photosynthesis

Summary

After studying this topic, you should be able to:
- describe the structure of a chloroplast
- describe the events of the light-dependent stage of photosynthesis, including the roles of the thylakoid membranes and the processes of cyclic and non-cyclic photophosphorylation
- describe the events of the light-independent stage of photosynthesis, including the roles of the stroma, RuBP, RUBISCO and reduced NADP and ATP in the Calvin cycle
- state that GALP is used as a raw material in the production of monosaccharides, amino acids and other molecules
- explain how carbon dioxide, light intensity and temperature can act as limiting factors for photosynthesis

■Topic 6 Microbiology and pathogens

Microbial techniques

Culturing microorganisms

Microorganisms, such as bacteria and fungi, can be grown in various **media** (singular: medium), including:

- nutrient broth — a liquid containing all the required nutrients in suitable concentrations
- nutrient agar — a jelly that you can make up from a powder or tablet
- a selective medium, a broth or agar preparation that allows only certain types of microorganisms to grow

All of these media must be **sterile**, to ensure that there are no organisms in the medium before you culture the desired microorganisms.

In a school laboratory, materials can be sterilised by heating to high temperatures, for example in an autoclave (pressure cooker), for long enough to kill all microorganisms and spores. This is suitable for glassware and liquids.

Aseptic techniques

All microbiological cultures should be treated as though they contain pathogens. It is also important to ensure that no unwanted microorganisms enter the cultures, and that no cultured microorganisms are able to spread into the surroundings. The techniques used are called **aseptic techniques**. See www.nuffieldfoundation.org/practical-biology/aseptic-techniques for full details.

When culturing organisms on nutrient or selective agar, the agar is heated to sterilise and liquefy it, before being poured into the plate. The aseptic techniques used for doing this are shown in Figure 20. Figure 21 shows a technique for inoculating the agar with bacteria.

The lid of the plate should then be taped down, to prevent it coming off accidentally and allowing other microorganisms to get in, or the ones already in there to get out. The tape should not go all around the dish, as this would produce anaerobic conditions.

The plate should then be incubated at a temperature below that of human body temperature, which would encourage the growth of pathogens. A suitable temperature might be 20–25°C.

The plate should be kept upside down, to prevent condensation dripping onto the surface of the agar.

See www.nuffieldfoundation.org/practical-biology/incubating-and-viewing-plates for more details.

> **Exam tip**
>
> Do not confuse the term 'aseptic' with 'antiseptic'.

1 Use sterile Petri dishes. Pre-sterilised plastic ones can be bought in packs; glass ones can be sterilised in an autoclave. Do not open the dishes until you are ready. Write labels on their bases.

2 Make up sterile, molten nutrient agar in a sterile conical flask. Remove the bung from the flask and place the bung in disinfectant.

3 Pass the opening of the flask through a Bunsen flame a few times, to kill any microorganisms that might have landed there.

4 Quickly lift the lid of a Petri dish, just enough to allow you to pour the agar in. Do not touch the inner surface of the lid. Pour the molten agar into the dish.

5 Cover the dish quickly and leave the agar for about 15 minutes, to cool and set.

Figure 20 Pouring a sterile agar plate

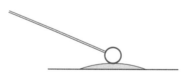

1 Hold an inoculating loop in a blue bunsen flame until it glows red hot, to sterilise it.

2 If using a culture grown in broth, dip the loop into the broth. If using a culture grown on agar, touch the top of the colony with the edge of the loop.

3 Lift the lid of the Petri dish just enough to allow you to put the loop inside. Make three streaks on the surface of the agar, using the edge of the loop.

4 Flame the loop again.

5 Make three more streaks as shown, spreading the microorganisms from the first three streaks at right angles.

6 Flame the loop again, and make three more streaks at right angles to the second three.

7 Flame the loop again, flip it over so that you are using the opposite edge, and make a final streak.

Note: this technique effectively dilutes the sample with each set of streaks, so that there should be somewhere on the plate where the bacteria grow as separate colonies.

Figure 21 Making a streak plate

Measuring the growth of a bacterial culture

Cell counts and dilution plating

This is normally used when the culture has been grown in a liquid medium. A **total cell count** includes all cells, including dead ones. A **viable count** includes only living cells.

A total cell count can be made using a **haemocytometer**. This is a special microscope slide in which a known volume of liquid is trapped beneath a cover slip. The slide is engraved with lines to form a grid (Figure 22). A sample is taken from the culture suspension, and placed on the slide. The slide is then viewed using the high-power objective lens of a microscope, and the number of cells visible within the grid is counted. Cells touching the top or left-hand boundaries are included in the count, whereas those touching the bottom or right-hand boundaries are not. If the known volume in which you have counted the cells on the haemocytometer is 0.004 mm^3, then you divide your count by 0.004 to estimate of the number of cells in 1 mm^3 of your sample.

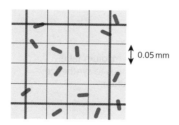

Figure 22 A haemocytometer grid

0.05 mm

Knowledge check 15

How many cells should be counted in the grid shown in Figure 22?

You can only count the cells on the haemocytometer grid if there are not too many or too few of them. For this reason, the normal procedure is to make a set of serial dilutions of the sample taken from the culture. One of these will provide the ideal number of bacteria on the grid squares — not too few, and not so many that the cells overlie one another and are impossible to count.

A viable cell count is made by taking several samples of known volume from the culture solution. The samples are diluted to give a range of different dilutions. A known volume of each dilution is then added to a nutrient agar plate. This is called **dilution plating**. After incubation, the number of colonies is counted. Each colony grew from a single bacterium, so the number of colonies is the same as the number of living bacteria in the sample that you added to the plate.

Optical methods

The total number of cells (both dead and alive) in a liquid culture can also be estimated using a colorimeter. The greater the number of cells in the culture, the more turbid it will be and the less light will pass through it. This technique is called **turbimetry**.

Measuring mass

A known volume of liquid can be taken from a liquid culture and centrifuged. The cells collect at the bottom of the centrifuge tube as a pellet, whose mass can be measured. This gives you wet mass. It is better to dry the pellet first, and measure dry mass. Like the colorimeter method, this does not tell you how many cells there are, and it does not distinguish between dead and living cells.

The bacterial growth curve

Figure 23 shows the stages of growth in a population of bacteria in a closed culture vessel.

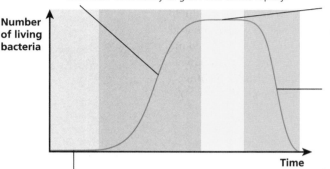

Log phase: Bacteria reproduce rapidly, with numbers doubling each generation; there are no factors limiting their growth other than their own ability to grow and divide rapidly.

Stationary phase: Population remains stable, as the rate of reproduction of new cells equals the rate of death of old cells; there are limiting factors that prevent more rapid reproduction, usually nutrients (lack of oxygen, change in pH and build-up of toxic metabolic products may also be limiting).

Death phase: Bacteria die much faster than new ones are being produced, so the population of living cells falls rapidly.

Lag phase: The bacteria that have been introduced to the medium take time to begin to reproduce; they may have to switch on genes to synthesize new enzymes to hydrolyse the nutrients present in the medium.

Figure 23 Bacterial growth curve

Calculating exponential growth rate constant

The growth rate constant, k, is a measure of how rapidly a population of bacteria is growing during the log (exponential) phase. The equation to use for this calculation is:

$$P_t = P_0 e^{kt}$$

where:

- P_t is the number of bacteria at the end of the time interval
- P_0 is the number of bacteria at the beginning of the time interval
- k is the growth rate constant
- t is the time interval

Imagine you measure the number of bacteria in a sample of fixed volume at 20 minutes, and find that it is 850. You measure the number again, in a sample of the same volume, at 50 minutes, and find that it is 4500.

$P_t = 4500$ $P_0 = 850$ $t = 30$ minutes $= 0.5$ hour

$4500 = 850e^{0.5k}$

$5.29 = e^{0.5k}$

$\ln 5.29 = 0.5k$

$1.67 = 0.5k$

$k = \dfrac{1.67}{0.5} = 3.34\,\text{hour}^{-1}$

Knowledge check 16

A bacterial culture contained 694 bacteria per mm^3 1 hour after inoculation, and 5752 bacteria per mm^3 4 hours later. Calculate the exponential growth rate constant.

Core practical 12

Investigating the rate of growth of bacteria in liquid culture

Use aseptic techniques throughout this investigation.

Make up a liquid medium, such as nutrient broth, suitable for the type of bacteria that you are going to grow. Place the medium in a sterile conical flask, plug with sterile cotton wool and allow to come to the temperature at which you wish to culture the bacteria.

Use a flamed inoculating loop to add a sample of the bacteria to the medium. Swirl gently to mix thoroughly. Use a sterile pipette to withdraw a small sample of the culture and use one of the methods described above (haemocytometer, dilution plating, turbimetry or mass) to estimate the number of bacteria in a known volume of sample. If possible, take three samples and calculate the mean number.

Take further samples from the culture at known time intervals and record the mean number of bacteria per unit volume. Plot this against time.

Core practical 13

Isolating individual species from a mixed culture of bacteria using streak plating

Use aseptic techniques throughout this investigation.

Make up a sterile nutrient agar plate. Make a streak plate using an inoculating loop loaded with a sample of the mixed culture, as described in Figure 21. Incubate until colonies of bacteria can be seen on the plate.

Use an inoculating loop to take a small sample from a single colony of bacteria and inoculate into nutrient broth, or onto nutrient agar.

Summary

After studying this topic, you should be able to:
- describe basic aseptic techniques for the safe handling of microorganisms
- describe the principles and techniques involved in culturing microorganisms, including the use of broth cultures, agar and selective media
- describe how to measure growth of a bacterial culture using cell counts, dilution plating, mass and turbimetry
- explain the different phases of a bacteria growth curve, and calculate the exponential growth rate constant
- describe how to investigate the rate of growth of bacteria in liquid culture, and how to isolate an individual species from a mixed culture using streak plating

Bacteria as pathogens

Some types of bacteria can cause infectious disease. An infectious disease is one that can be passed between one person and another. Microorganisms that cause disease are known as **pathogens**.

Pathogenic bacteria cause illness by:

- invading host tissues and damaging cells, for example *Mycobacterium tuberculosis* causes tuberculosis (TB) by entering and destroying cells in the lungs and other tissues
- producing harmful chemicals known as **toxins**, for example *Staphylococcus* spp. and *Salmonella* spp.

Staphylococcus releases its toxins into the surrounding tissues. They are therefore said to be **exotoxins**. These have widespread effects in the body and can cause toxic shock syndrome.

Salmonella toxins are part of the cell wall of the bacterium; they are not released into the body. They are called **endotoxins**.

Exam tip

Remember that a bacterium is an organism — a complete cell. Toxins are just molecules.

Summary

After studying this topic, you should be able to:
- state that pathogenic bacteria can be agents of infection
- state that *Mycobacterium tuberculosis* invades host tissue and destroys it
- state that *Staphylococcus* produces exotoxins, and *Salmonella* produces endotoxins

Action of antibiotics

An **antibiotic** is a substance that, when taken orally or by injection, kills or inhibits the growth of bacteria but does not harm human cells. Antibiotics are not effective against viruses.

Antibiotics act on structures or metabolic pathways that are found in bacteria but not in eukaryotic cells. The correct antibiotic must be chosen for a particular disease. For example:

- **penicillin** prevents the synthesis of the links between peptidoglycan molecules in bacterial cell walls; when the bacteria take up water by osmosis, the cell wall is not strong enough to prevent them bursting
- **tetracycline** binds to bacterial ribosomes and inhibits protein synthesis

Penicillin is a **bactericidal** antibiotic, meaning that it kills bacteria. Tetracycline is **bacteriostatic**, meaning that it does not directly kill bacteria, but prevents them from growing and reproducing.

Knowledge check 17

Explain why neither penicillin nor tetracycline has any effect on viruses.

Summary

After studying this topic, you should be able to:
- explain how penicillin and tetracycline act as antibiotics
- explain the difference between bactericidal and bacteriostatic antibiotics

Antibiotic resistance

Exposure to antibiotics exerts strong selection pressure on bacterial populations. Any bacterium that is **resistant** to the antibiotic — for example, because it has an allele of a gene that causes the synthesis of an enzyme that can break down the antibiotic — has a selective advantage and is more likely to survive and reproduce successfully. The offspring will inherit the alleles that confer resistance. A whole population of resistant bacteria can therefore be produced.

Alleles that confer resistance are often found on plasmids in bacteria. Plasmids can be passed from one bacterium to another, and even between different species of bacteria. In this way, antibiotic resistance can be transferred from one species to another.

To reduce the risk of antibiotic resistance developing and spreading, it is important that:

- antibiotics are only used when necessary — for example, they should not be used to treat viral diseases or diseases that will get better without treatment, or be given to livestock on a regular basis to increase growth rates
- a person prescribed antibiotics should complete the course, as this increases the chances of eradicating all the disease-causing bacteria in the body
- consideration is given to using more than one antibiotic to treat a disease, because it is unlikely that any one bacterium will possess two different resistance alleles
- hospitals and health centres take great care not to spread bacteria from one patient to another, by ensuring that staff follow strict hygiene regulations, such as washing hands thoroughly after touching a patient or anything with which a patient has been in contact, and using bactericidal hand gels regularly

Exam tip

Avoid saying that bacteria 'evolve so that they become' able to resist the immune response, as this implies that they change purposefully.

Summary

After studying this topic, you should be able to:
- explain how antibiotic resistance develops and spreads in bacteria
- explain the methods and difficulties of controlling the spread of antibiotic resistance in bacteria

Other pathogenic agents

Fungi, viruses and protoctists can also be pathogens.

Stem rust on cereal crops is caused by fungi — for example *Puccinia graminis* infects wheat. The fungus spreads as spores carried by wind, or by direct contact with other infected wheat plants. The spores germinate when they land on the stem of the plant.

The fungus then grows hyphae that invade the wheat tissue. The wheat is harmed because the hyphae absorb nutrients, damage vascular tissue and weaken the stem, making it more likely that the plant will fall over during strong winds or heavy rain.

Influenza is caused by a virus. The virus is spread in droplets of moisture exhaled by an infected person; it can also survive on hands and other surfaces for some time and can therefore be transmitted by touching a surface that an infected person has touched, and then bringing the hands to the face so that the virus can enter the mouth or nose.

The virus inserts its RNA into a host cell in the respiratory passages or lungs, where the RNA is used to make DNA. The code carried by this DNA is then used by the host cell to make new viruses, which burst out of the cell and destroy it. The virus causes fever, headaches and fatigue, which are the result of tissue damage. It is also possible that these symptoms are caused by the person's immune response to the virus (page 36), which produces widespread inflammation.

Malaria is caused by a protoctist, *Plasmodium*. There are several species, which cause different types of malaria. In a person the *Plasmodium* infects red blood cells and breeds inside them. Toxins are released when the *Plasmodium* pathogens burst out of the cells, causing fever.

Plasmodium is transmitted in the saliva of female *Anopheles* mosquitoes, which inject saliva to prevent blood clotting when they feed on blood from a person. When a mosquito bites an infected person, *Plasmodium* is taken up into the mosquito's body and eventually reaches its salivary glands. The mosquito is said to be a **vector** for malaria.

Knowledge check 18

Explain the difference between the cause of a disease and a vector for a disease.

Exam tip

Note that the same term — vector — is used for an organism that transfers a pathogenic organism from one person to another, and for a method of transferring genes from one cell to another in gene technology.

Summary

After studying this topic, you should be able to:
- describe and explain the transmission, mode of infection and pathogenic effect of:
- stem rust fungus on cereal crops
- the influenza virus
- the malarial parasite

Problems of controlling endemic diseases

In some countries, malaria is an **endemic** disease, meaning that it is always present in the population.

Various measures are used to control endemic malaria.

- Sources of water in which mosquitoes can breed can be removed, or the water can be sprayed with insecticides to kill larvae. This can harm other organisms that are not harmful, or that may be beneficial (e.g. pollinators such as bees). Some insecticides can be harmful to humans.
- Large numbers of sterile male mosquitoes can be released. This reduces the chances of a female mosquito mating with a fertile male and therefore reduces the size of the next generation of mosquitoes. It can be difficult for local people to understand the value of releasing large numbers of mosquitoes. Some such projects use genetically modified males, which can cause concern.
- Sleeping under a mosquito net, or wearing long-sleeved clothing and using insect repellant, can reduce the chances of a mosquito picking up *Plasmodium* from an infected person, or passing it to an uninfected person.
- Prophylactic drugs (that is, drugs that prevent pathogens infecting and breeding in a person) can be taken. However, in many parts of the world *Plasmodium* has evolved resistance to some of these drugs. The drugs can also be expensive.
- Vaccines against malaria have largely been unsuccessful, but new types of vaccine now show promise in reducing the chance of infection in children.

Exam tip

Remember that only female mosquitoes bite, and therefore there is no danger of malaria being transmitted by male mosquitoes.

In general, it is important to use several different methods in order to try to eliminate malaria from a country. In order to evaluate the usefulness of any particular method, data should be collected that allow comparison between the rates of infection with and without a particular method being put into place. Scientists can then use these data to determine the effect, if any, of the control measure. However, it is unethical to deny potential malarial control to some people on the grounds of doing research.

Summary

After studying this topic, you should be able to:
- explain what is meant by an endemic disease
- describe some control methods for endemic malaria
- outline some social, economic and ethical implications of these control methods

Response to infection

The human immune system is made up of the organs and tissues involved in destroying pathogens inside the body. There are two main groups of cells involved:
- **phagocytes**, which ingest and digest pathogens or infected cells
- **lymphocytes**, which recognise specific pathogens through interaction with receptors in their cell surface membranes, and respond in one of several ways, for example by secreting antibodies

Phagocytes include macrophages and neutrophils (Figure 24). They engulf bacteria by endocytosis and digest them inside phagosomes. Macrophages can live for several months, whereas neutrophils generally live for only a few days.

Macrophage   Neutrophil

Figure 24 A macrophage and a neutrophil

Lymphocytes are of two main types — B cells and T cells. They look identical, but have different functions. B cells are largely involved with the humoral response, and T cells with the cell-mediated response. However, these two types of response are not clearly divided, and there is much interaction between the B cells and T cells.

The humoral immune response

All our cells contain genes that are able to produce a very wide range of proteins called **immunoglobulins** (Figure 25).

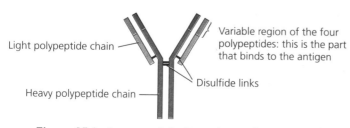

Light polypeptide chain

Variable region of the four polypeptides: this is the part that binds to the antigen

Disulfide links

Heavy polypeptide chain

Figure 25 An immunoglobulin molecule (an antibody)

Exam tip

Take care not to confuse antigens, antibodies and antibiotics.

These can act as **antibodies**, which attach to, and result in the destruction of, bacteria. We have thousands of different B cells, in each of which a different type of antibody can be produced. This is possible because of post-transcriptional processing of the genes that code for immunoglobulins, producing many different versions of mRNA and therefore many different versions of the proteins.

B cells place some of these antibodies in their cell surface membranes, where they act as receptors. The variable regions of the antibodies have a complementary shape to the antigens on pathogens against which they act. When this antigen contacts the B cell receptor, the B cell is activated. It divides repeatedly by mitosis, producing a clone of cells. This is called **clonal selection** (Figure 26).

Some of these develop into **plasma cells**, which synthesise and secrete large amounts of the antibody. Their activity is so intense that they do not usually live for more than a few weeks, but they are replaced by more plasma cells if need be. Others remain in the blood but do not secrete antibody. They are **memory cells**, and they live for a very long time. Their continued presence in the blood, perhaps for many years after the original infection, means that the immune system can mount an instant attack on the same pathogen should it invade the body again. The person has become immune to that particular disease.

Knowledge check 19

Plasma cells contain large numbers of mitochondria, extensive rough endoplasmic reticulum and several Golgi bodies. Explain why.

Exam tip

Antigen-presenting cells are sometimes called APCs — but it is best if you write their full name in an exam answer.

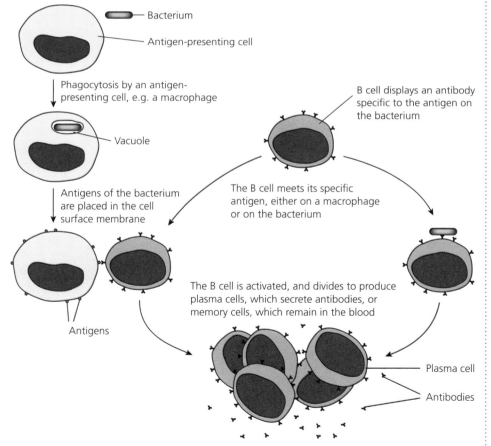

Figure 26 The response of B cells to an antigen

B cells may initially come into contact with the antigen in the blood plasma or body fluids, or they may meet it on an **antigen-presenting cell**. Several different types of cell, including macrophages, act as antigen-presenting cells. They place antigens of pathogens they have encountered in their cell surface membranes, where there is a good chance that a B cell will encounter them.

The cell-mediated immune response

T cells, like B cells, place receptors in their cell surface membranes, and these bind specifically with antigens. However, T cells will only respond to antigens if they find them in the cell surface membranes of body cells. These may be antigen-presenting cells such as macrophages, or they may be cells that have been infected by viruses and have placed molecules from the virus in their membranes as a 'help' signal. T cells respond in a similar way to B cells when they meet their antigen, quickly producing a clone (Figure 27). Some of these remain inactive, lasting for many years as T memory cells. Others become **T helper cells**, which secrete **cytokines** (for example **interferon**) that stimulate other cells, such as macrophages and B cells, to become active against the virus. Yet others become **T killer cells**, which destroy the cells that display the antigen from the infective agent in their cell surface membranes.

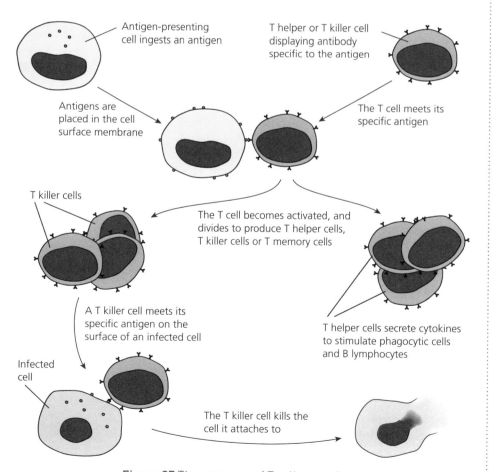

Figure 27 The response of T cells to antigens

Exam tip

Do not confuse cytokines (produced by T cells) with antibodies (produced by B cells).

T cells are therefore especially important in the immune response against viruses, and also against our own body cells if these are malfunctioning, such as becoming cancerous. T cells are also involved in the rejection of transplanted tissue whose antigens do not closely match those of the recipient. It is one particular group of T cells that is put out of action by the human immunodeficiency virus, HIV, which causes AIDS.

The development of immunity

Active immunity

After a person has survived an infection, memory B and T cells remain in the blood. If the same antigen is encountered on a second occasion, these memory cells enable a faster, more intense immune response to be mounted. More antibodies are produced, more quickly. This is called the **secondary immune response**. It may destroy the pathogen before any symptoms of infection are shown.

Active immunity occurs when the person has made their own antibodies against a pathogen, either as the result of an infection as described above (**natural immunity**) or after being vaccinated with a weakened form of the pathogen (**artificial immunity**). Memory cells remain in the body for many years, so this type of immunity tends to be very long-lasting.

Passive immunity

Passive immunity occurs when ready-made antibodies enter the person's body. A fetus acquires antibodies from its mother while in the uterus, and there are also antibodies in breast milk. This is a form of natural immunity.

A person can also obtain antibodies by injection. This may be done if they are thought to have already been infected with a serious disease, and need instant help to destroy the pathogens. Antibodies may also be given to people travelling to an area where there is a high risk of infection, such as aid workers going to earthquake zones. This is a form of artificial immunity.

Passive immunity generally lasts only weeks or months at best, because the person does not have their own memory cells, and the antibodies they have been given do not last long.

Vaccination programmes

A vaccine contains a weakened (attenuated) form of a pathogen or toxin. It is injected into the body, or taken orally, and acts as an antigen that brings about an immune response. The resulting memory B and T cells provide immunity. In some cases, more than one dose of vaccine is required to increase immunity to a high level.

Exam tip

Most vaccines cannot be taken orally, because enzymes and hydrochloric acid in the alimentary canal destroy the active ingredients. The polio vaccine is an example of one that is given orally; it contains attenuated viruses that cause an immune response in the cells lining the small intestine.

Knowledge check 20

A person cuts their hand on a dirty gardening spade, and is given an injection containing antibodies against the tetanus bacterium. Explain why they were given antibodies rather than a vaccination containing weakened bacteria, and state the type of immunity they will now have.

Most countries have vaccination programmes in which young children are given a series of vaccinations against potentially serious infectious diseases.

If a large proportion of the population is vaccinated, this confers **herd immunity**. As only the unvaccinated people are able to harbour the pathogen, this makes it unlikely that the pathogen will be encountered by unvaccinated people (e.g. very young children) or those with compromised immune systems (e.g. very elderly people, or those with HIV).

If, however, some parents do not allow their children to be vaccinated, then all other unprotected people have a much greater chance of becoming infected and possibly dying from the disease.

Summary

After studying this topic, you should be able to:

- describe the action of macrophages, neutrophils and lymphocytes
- explain the humoral immune response
- explain the cell-mediated immune response
- explain how immunity develops

- understand the differences between natural, artificial, active and passive immunity
- describe how vaccination can be used to control disease, the importance of herd immunity, and the problems that arise if not all individuals are vaccinated

■ Topic 7 Modern genetics

Using gene sequencing

The complete set of genes that is present in an organism's cells is known as its **genome**.

The polymerase chain reaction

It is now possible to find the sequence of bases in the genome of a person (or other organism). Even tiny samples of DNA can be used, because they can be amplified using the **polymerase chain reaction** (**PCR**). This is an automated method of making multiple copies of a sample of DNA. The DNA is exposed to a repeating sequence of different temperatures, allowing different enzymes to work. The process shown in Figure 28 is repeated over and over again, eventually making a very large number of identical copies of the original DNA molecule.

The primer is a short length of DNA with a base sequence that is complementary to the start of the DNA strand to be copied. This is needed to make the DNA polymerase begin to link nucleotides together as it makes a copy of the exposed DNA strand.

Making use of gene sequences

Once the base sequence of a DNA sample is known, it can be used to predict the amino acids coded for by the DNA triplets in a gene, and therefore to predict the proteins that will be formed. This can provide information about whether or not a person will develop a genetically determined condition, such as Huntington's disease or cystic fibrosis.

Gene sequencing is also used in forensics. For example, a small sample of DNA found at the scene of a crime can be amplified, and then sequenced. Particular regions of the genome are especially variable, and the base sequences of these regions are only likely to be identical in a tiny number of cases.

DNA samples from the crime scene can be matched against the sequences of people known to be at the scene and also any suspects. This can be done by placing the DNA sample on a gel across which a voltage is applied (electrophoresis — described on page 11 of the second student guide in this series). The electrical charge causes the DNA sections to move along the gel, and lengths of DNA with the same base sequences end up at the same position on the gel. If the sequences of two samples match perfectly, there is a high chance that they came from the same person.

DNA profiling can also be used to determine the father of a child (paternity testing — Figure 29). The base sequences on the child's DNA have all been inherited from their mother and father. If a particular base sequence in the child's DNA does not match its mother's, then this base sequence will be found in the father's DNA. If a man thought to be the father of the child does not have this base sequence, then he is not the biological father of the child.

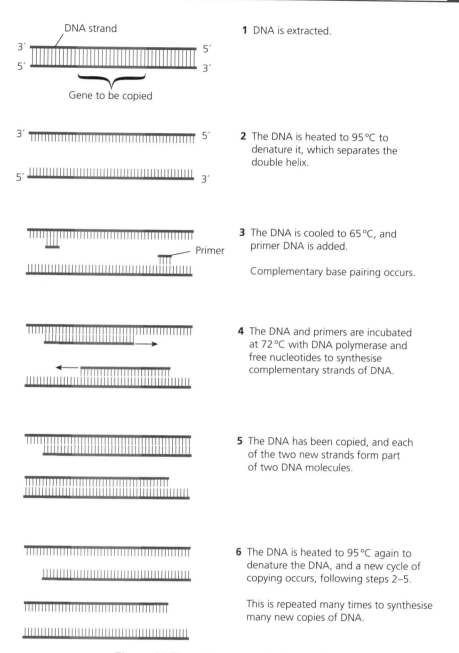

1 DNA is extracted.

2 The DNA is heated to 95 °C to denature it, which separates the double helix.

3 The DNA is cooled to 65 °C, and primer DNA is added.

  Complementary base pairing occurs.

4 The DNA and primers are incubated at 72 °C with DNA polymerase and free nucleotides to synthesise complementary strands of DNA.

5 The DNA has been copied, and each of the two new strands form part of two DNA molecules.

6 The DNA is heated to 95 °C again to denature the DNA, and a new cycle of copying occurs, following steps 2–5.

  This is repeated many times to synthesise many new copies of DNA.

Figure 28 The polymerase chain reaction

DNA samples from two mothers, their children and possible fathers were separated using electrophoresis. The diagrams show the positions at which lengths of DNA came to rest on the gel.

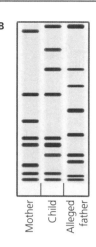

Figure 29 Paternity testing

Exam tip

DNA sequencing can also be used to determine how closely related two species of plant or animal are. The more similar the base sequences in their DNA, the more closely they are related.

Knowledge check 21

In the sets of results shown in diagrams A and B in Figure 29, could the alleged father be the actual biological father of the child? Explain both of your answers.

Summary

After studying this topic, you should be able to:
- explain what is meant by the term genome
- explain how PCR is used to amplify DNA samples
- explain how DNA samples can be used to predict amino acid sequences of proteins and possible links to genetically determined conditions
- explain how DNA sequencing can be used in forensics to identify criminals and for paternity testing

Factors affecting gene expression

All of the cells in your body contain a full set of genes. However, in each cell only a certain set of these genes is used to make proteins (expressed). Different genes are expressed in different cells, and at different times.

Transcription factors

A **transcription factor** is a protein molecule that binds to DNA and either causes or prevents the transcription of a gene. Transcription factors do this by either making it easier for the enzymes involved in transcription (including RNA polymerase) to bind to the DNA, or preventing them from doing so.

An example of a transcription factor in the seeds of cereal plants is phytochrome-interacting factor (**PIF**).

- PIF can bind to a region of DNA called a **promoter**. This allows the gene coding for amylase to be transcribed (Figure 30).
- Amylase is only required when the seed is germinating, so the gene is only 'switched on' at this time. At other times, PIF is bound to a protein called DELLA, which stops it binding with the promoter.

Exam tip

A promoter is a length of DNA to which RNA polymerase can bind, allowing the RNA polymerase to begin to transcribe the DNA.

- When the seed is about to germinate, the plant growth substance gibberellin is secreted by the seed. The gibberellin binds with an enzyme that is then able to break down the DELLA protein.
- This allows PIF to bind with the promoter, so that amylase synthesis can begin as the seed germinates.

1 PIF is a transcription factor that is normally bound to a DELLA protein.

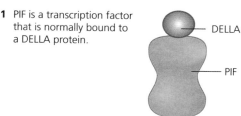

2 When the plant growth factor gibberellin is present, it binds with a receptor and an enzyme. This complex breaks down the DELLA protein.

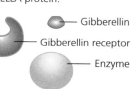

3 PIF can now bind with the promoter for the amylase gene, so mRNA for amylase production is produced by transcription.

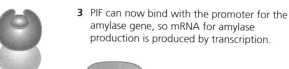
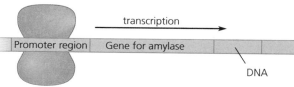

Figure 30 How the transcription factor PIF works

Post-transcriptional modification of mRNA

The mRNA transcribed from a particular gene is not all used to synthesise a protein. The parts of the gene that are used to make a protein are called **exons**, and the parts that are not are called **introns**.

Transcription, however, takes place for the whole length of the gene, so the first mRNA molecule that is produced contains a complementary sequence for the introns as well as the exons. It is sometimes known as **pre-mRNA**. The intron sections are then cut away, and the exons joined together. This is called **mRNA splicing** (Figure 31).

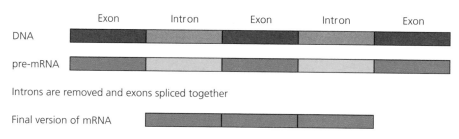

Figure 31 Post-transcriptional mRNA splicing

The final version of the mRNA molecule, now containing only the code from the exons, leaves the nucleus and travels to a ribosome.

The pre-mRNA molecules can be spliced together in different ways. This means that one gene can produce several different versions of mRNA, which, of course, means that different sequences of amino acids are coded for. Thus one gene can code for several different proteins. An example is the production of a huge number of different types of antibody from a relatively small number of different genes.

Epigenetic modification

Epigenetics involves genetic control by factors other than the sequence of bases in an individual's DNA.

DNA methylation

Methyl groups can be added to particular DNA sequences (Figure 32). The insertion of a methyl group changes the structure of the DNA, and therefore affects the way that the gene interacts with other molecules that are involved with initiating and carrying out transcription. Methylation therefore prevents the gene being expressed. The gene is said to be 'silenced'.

Histone modification

In a chromosome, DNA is closely associated with proteins called **histones**. DNA is wound around the histones, producing the chromatin that is visible in the nucleus of a non-dividing cell. When the DNA–histone complex is tightly condensed (heterochromatin) the DNA cannot be transcribed. The shape and behaviour of histones can be altered by adding acetyl groups or methyl groups, which affects whether the DNA is condensed or not (Figure 32). This can determine whether or not the DNA can be transcribed.

Exam tip

In a transmission electron micrograph, heterochromatin is darkly stained, while euchromatin, which is much less tightly coiled, looks paler.

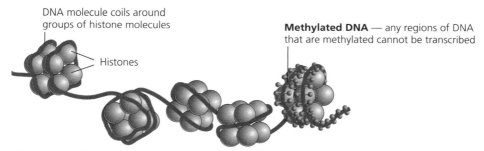

DNA molecule coils around groups of histone molecules

Histones

Methylated DNA — any regions of DNA that are methylated cannot be transcribed

If the DNA and histones are very tightly coiled (condensed), the DNA cannot be transcribed.

Figure 32 DNA, histones and methylation

RNA-associated silencing

Small lengths of RNA, which do not code for protein synthesis, can bind with complementary base sequences on DNA and prevent them from being used in transcription.

Significance of epigenetic modification

Epigenetic modifications are essential for **cell differentiation**, ensuring that only particular sets of genes are expressed in specific types of specialised cell.

Epigenetic modifications are long-lasting. They can be passed on from parent to offspring.

Summary

After studying this topic, you should be able to:

- explain what transcription factors are, and how they affect gene expression
- explain how RNA splicing can result in different products from a single gene
- explain how DNA methylation, histone modification and non-coding RNA are involved in epigenetic modification
- outline the importance of epigenetic modification in cell differentiation

Stem cells

Like all animals, you began your life as a single cell — a zygote — formed by the fusion of a male and female gamete. The zygote divides by mitosis to form a ball of cells. Each of these cells has the capability of differentiating to form any of the specialised cells that make up a human body. They are said to be **totipotent stem cells**.

A **stem cell** is a cell that has not differentiated, and retains the ability to divide to form other cells that can differentiate into several kinds of specialised cell. Only the cells that are formed in the first two divisions of the zygote are totipotent, meaning that they can give rise to a complete organism.

As these totipotent stem cells continue to divide, they form a ball of cells called a **blastocyst**. Further cell division forms an **embryo**, in which the cells are **pluripotent**. Pluripotent stem cells are also able to produce all kinds of specialised cells, but they cannot form a whole organism.

As the embryo continues to develop, most cells become differentiated, and are no longer able to form other types of cell. However, some stem cells remain, and these become adult stem cells. These are **multipotent**. Multipotent stem cells are able to form only a limited range of specialised cells. For example, stem cells in the bone marrow can divide to form all the different types of blood cell, but they cannot form neurons or skin cells. All of the stem cells found in adults are multipotent.

Using stem cells in medicine

Pluripotent stem cells derived from early embryos have the potential to be used in the treatment of many different diseases. For example, Parkinson's disease is caused by the loss of neurons in the brain that secrete dopamine. Stem cells from embryos have been transplanted into the brains of Parkinson's disease patients. Initial trials carried out in the late 1990s seemed to show little success, but some of those patients eventually showed great improvements in their condition 20 years after the transplants. This suggested that new dopamine-producing cells were produced, but took time to 'bed in'. New trials are now underway.

There are ethical issues in obtaining stem cells from embryos:

- Stem cells can be obtained from 'spare' embryos produced for in vitro fertilisation. These embryos could potentially have developed into babies.
- Stem cells can be obtained by taking a single cell from a very early embryo, without destroying the embryo. There is considerable evidence that this still allows the embryo to develop normally, but concerns remain that this may not always be true.

Knowledge check 23

Suggest what happens in the cells to change the totipotent cells to pluripotent cells in an early embryo.

Reprogramming stem cells

The change of totipotent stem cells into pluripotent and multipotent stem cells is at least partly caused by epigenetic modifications. Research is ongoing to find ways of reversing these changes, so that multipotent stem cells from adults could be converted back to pluripotent or totipotent stem cells that could have uses in medicine.

For example, fibroblasts are adult stem cells found in the connective tissues in the body. Researchers have reprogrammed fibroblasts by introducing particular genes that encode transcription factors that allow genes that had been switched off to be expressed again. In the earliest successful trials, the four genes used were Oct4, Sox2, c-MYC and Klf4. Since then, different genes have also shown success. The new cells produced are called **induced pluripotent stem cells**, or **iPS** cells. These iPS cells can form other cells such as liver cells. These could be used to treat people with liver disease.

Using reprogrammed cells raises fewer ethical issues than using embryonic stem cells, as they are obtained from adult tissues.

Summary

After studying this topic, you should be able to:
- explain what is meant by a stem cell, and explain the difference between totipotent, pluripotent and multipotent stem cells
- describe some examples of the potential use of stem cells to treat medical conditions
- discuss the ethical issues in using stem cells obtained from embryos
- explain how epigenetic modifications are involved in the change of totipotent stem cells to pluripotent and multipotent cells, and how reversing these changes can form induced pluripotent stem cells

Gene technology

Gene technology, or genetic engineering, is the manipulation of genes in living organisms.

Genes extracted from one organism can be inserted into another. This can be done within the same species (for example, in gene therapy) or genes may be transferred from one species to another.

Steps in producing recombinant cells

DNA that has been produced by linking together DNA from two or more different sources is said to be **recombinant**. A cell or organism containing recombinant DNA is said to be a recombinant cell, or **genetically modified**. If the DNA has come from a different species, it is also known as a **transgenic organism**.

Obtaining the required DNA

There are three main ways of obtaining the required length of DNA to be transferred:
- The required length of DNA (gene) is cut out using enzymes called **restriction endonucleases**. There are many different restriction endonucleases. Each one cuts DNA at a particular base sequence. They often cut the DNA at different points in the two strands, leaving a section of unpaired DNA called a **sticky end**.
- Genes can also be synthesised in the laboratory by linking DNA nucleotides together in a particular order.

■ Genes can be synthesised by isolating mRNA that has been transcribed from the required gene. The mRNA is then incubated with the enzyme **reverse transcriptase**, which makes a length of DNA that is complementary to the mRNA.

Exam tip

Take care that you understand the difference between restriction endonucleases and reverse transcriptase.

Knowledge check 24

When bacteria were first engineered to produce human insulin, the gene for insulin was obtained by extracting mRNA from beta cells in the pancreas, and incubating it with reverse transcriptase. Suggest the advantages of using this mRNA as a starting point, rather than simply extracting DNA from the cells.

Inserting the required DNA into cells

In gene technology, a **vector** is an organism or structure that inserts the recombinant DNA into the required cell. Four methods of inserting recombinant DNA are described below.

■ **Plasmids** are small, circular DNA molecules found in many bacteria. A plasmid can be cut open using the same restriction endonuclease that was used to cut out the required DNA from its original source. This leaves sticky ends with base sequences complementary to those on the DNA. The plasmids are then mixed with the required DNA, allowing the bases on the plasmids to form hydrogen bonds with the bases on the required DNA. The sugar–phosphate backbones of the DNA molecules are then joined using the enzyme **DNA ligase**, 'closing the circle' on the plasmids (Figure 33).

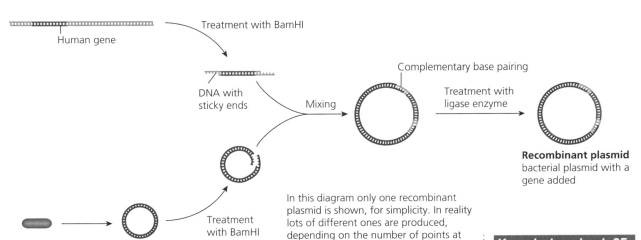

Figure 33 One method of producing recombinant DNA

The plasmids can then be mixed with bacteria, and calcium ions added to make the bacterial cells more able to take up plasmids. A small proportion of the bacteria will take up a plasmid, some of which will be recombinant plasmids. This produces transformed bacterial cells.

■ **Viruses** can be used as vectors. Genes in the viruses that can cause harm are removed, and are replaced with the desired gene. The virus is then allowed to infect the cells that are to be transformed.

Knowledge check 25

When the 'broken' plasmids have reformed into complete circles after treatment with DNA ligase, would you expect all of them to contain the required DNA? Explain your answer.

If the virus is a DNA virus, it inserts its DNA into a cell, including the recombinant DNA. Retroviruses, which contain RNA and not DNA, are also often used as vectors. They are genetically engineered to contain RNA with a complementary base sequence to the required gene. They insert their RNA into the cells, where the cell uses it to make DNA.

- The bacterium **Agrobacterium tumefaciens** is often used as a vector for introducing recombinant DNA to plant cells. This bacterium naturally infects plant cells, which it does by inserting part of a plasmid into the plant cells' DNA. This causes the plant cells to divide to form a tumour, so the plasmid is known as a tumour-inducing plasmid or Ti plasmid. The Ti plasmid can be engineered as described above, and inserted into A. tumefaciens, which can then used as a vector to introduce a desired gene into the plant cell.

- **Gene guns** bombard cells with minute particles of gold, which have been coated with many copies of the desired DNA. Some of these gold particles will enter cells, which may take up the introduced DNA and integrate it into their genomes.

Identifying transformed cells

Only a small percentage — perhaps 1% — of the cells treated as described in the previous section will actually take up the desired DNA and incorporate it into their genomes. It is therefore necessary to identify these recombinant cells, so that they can be separated from the others.

In the early days of gene technology, genetically modified bacteria were often identified using **antibiotic resistance genes**. Plasmids often contain genes for resistance to antibiotics. For example, the plasmid pBR322 contains genes that confer resistance to tetracycline and ampicillin.

1 The plasmid is cut open using the restriction enzyme BamH1, which cuts at a base sequence in the middle of the tetracyline resistance gene. The desired DNA is then inserted into this point. The sticky ends of the broken plasmids and desired DNA are then linked up using DNA ligase. Not all of the plasmids will be transformed — some will just join up again without the desired DNA in place.

2 The plasmids are then mixed with the bacteria to be transformed. Some of the bacteria will take up plasmids, some of which will be transformed plasmids, and some of which will not.

3 All of the bacteria that have taken up the plasmids will be able to grow on agar jelly containing the antibiotic ampicillin. The bacteria that have taken up untransformed plasmids (in which both antibiotic resistance genes are intact) will also be able to grow on agar containing tetracycline. But the bacteria that have taken up the transformed plasmids in which the tetracycline gene has been split by the inserted DNA will not be resistant to tetracycline. These transformed bacteria are identified using **replica plating** (Figure 34).

Today, antibiotic resistance genes are less widely used as markers to identify transformed bacteria, because of the concern of increasing the risk of pathogenic bacteria becoming resistant to antibiotics. Instead, genes such as those for an enzyme that produces green fluorescent protein, GFP, are often used. The gene for this enzyme is inserted into the plasmids along with the desired gene. Any cells that have taken up the transformed plasmids can be identified by shining ultraviolet light on them, which causes them to glow green.

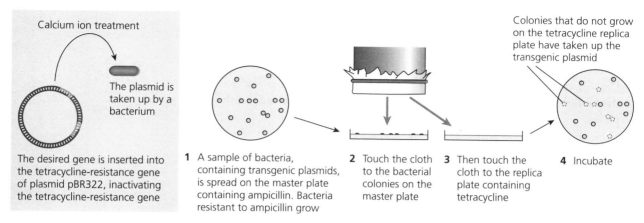

Figure 34 Identifying transformed bacteria by replica plating

Examples of genetically modified organisms

Knockout mice

Researchers can produced genetically modified mice in which a particular length of DNA (gene) has been inactivated. This can be done by inserting extra DNA into the genome of the mouse. Normally this is done in embryonic stem cells, so that all of the cells in the adult mouse that develops from these cells lack this gene.

This procedure can be used to determine the functions of different genes, by looking at the phenotypes of the mice in which the gene has been 'knocked out'. For example, it has been found that knocking out a gene for a protein called p53 greatly increases the risk that the mouse will develop cancer, indicating that this gene is involved in the control of the mitotic cell cycle.

Genetically modified soya beans

Soya beans are a major crop in many parts of the world — they are a high-protein, high-lipid food. They are used in the manufacture of a wide range of food products.

Soya beans have been genetically modified in several different ways, for example:

- to be resistant to herbicides such as glyphosate
- to contain a protein called Bt, obtained from bacteria, which is toxic to certain types of insect; this makes the plants resistant to insect pests, so that insects feeding on the beans are killed
- to reduce the percentage of linoleic acid in lipids, as this readily oxidises and makes the oil and products made from it taste rancid; instead, the beans produce lipids with more oleic acid and/or stearic acid.

Exam tip

Remember that the correct term to use here is 'resistant' not 'immune'.

Debate about the use of genetically modified organisms

There is a wide-ranging debate about the use of genetically modified organisms (GMOs). Many of the concerns raised by people stem from a lack of understanding of what a GMO is. However, there are also some valid concerns that need to be taken into consideration. A few of these issues are outlined below, but you will find many more examples in the media.

Content Guidance

Potential benefits of using genetically modified organisms include the following:

■ Crop plants can be produced that are resistant to attack by certain pests, reducing the need to use pesticides. This can increase yields and reduce the risk of harming beneficial insects.

■ Crop plants can be produced that are resistant to a particular herbicide, allowing the herbicide to be sprayed on the crop where it will kill weeds but not the crop plants.

■ Crop plants can be modified so that they produce higher quantities of a particular nutrient, for example Golden Rice, which produces beta-carotene (a precursor of vitamin A) and therefore reduces the risk of children suffering from vitamin A deficiency.

■ Proteins, including drugs such as insulin, can be produced by genetically modified bacteria. This avoids harvesting the proteins from an organism (e.g. in the past insulin was obtained from the pancreases of pigs) or producing them by chemical synthesis (which can be very expensive).

Potential risks and concerns include the following:

■ Genes inserted into a crop plant might spread to others via pollen. This could cause changes in the genotypes of wild plant populations, which could adversely affect other organisms in the ecosystem. For example, genes from crops engineered to be resistant to insects could spread into wild plants, so that insects would no longer have a good food supply. There has so far been limited evidence that this has ever occurred, but it is a genuine risk.

■ Pests might develop resistance, through natural selection, to the substance in GM crops that confers resistance to pests; this could result in a population of 'superpests'. This is a genuine risk, but it is probably no greater than the risk associated with pests developing resistance to pesticides.

■ Some people consider that consuming foods made from, or containing, GM organisms could be harmful to health. There is no evidence that this is a genuine risk.

■ Seed for GM crops is more expensive for farmers to buy than non-GM seed. This can put economic pressure on poor farmers in developing countries who cannot afford GM seed, causing great stress and hardship.

Summary

After studying this topic, you should be able to:
■ describe how recombinant DNA can be produced, including the use of restriction endonucleases and DNA ligase
■ describe the use of plasmids, viruses, *Agrobacterium tumefaciens* and gene guns for the introduction of recombinant DNA to cells
■ explain how antibiotic resistance genes and replica plating can be used to identify successfully transformed cells
■ explain how knockout mice can be used as animal models to investigate gene function
■ explain how genetic modification of soya beans can improve production and reduce the tendency for products to undergo oxidation
■ discuss some of the issues relevant to the debate concerning the widespread use of genetically modified organisms

Questions & Answers

In this section there are two sample examination papers, which contain questions in styles similar to those in the Edexcel Advanced GCE papers 1 and 3 of the Biology B specification. However, whereas the Advanced GCE paper 3 tests content from all or most of the topics you will study during your course, these sample papers test only content from Topics 1 to 7.

You have 1 hour 45 minutes to answer paper 1, and 2 hours 30 minutes to answer paper 3. There are 90 marks on paper 1 and 120 marks on paper 2, so you can spend around 1 minute per mark, plus time for reading the questions and checking your answers. If you find you are spending too long on one question, move on to another that you can answer more quickly. If you have time at the end, come back to the difficult one.

Some of the questions require you to recall information that you have learned. Be guided by the number of marks awarded to suggest how much detail you should give in your answer. The more marks there are, the more information you need to give.

Some of the questions require you to use your knowledge and understanding in new situations. Don't be surprised to find something completely new in a question — something you have not seen before. Just think carefully about it, and find something that you do know that will help you to answer it.

Some questions require you to carry out calculations. It is important that you show all of your working clearly, step by step. There are often marks for the working, and even if you end up with the wrong answer, you can still get marks for having carried out a particular step correctly. Paper 3 is particularly likely to have calculations as part of a question, but calculations can also appear in paper 1.

Think carefully before you begin to write. The best answers are short and relevant — if you target your answer well, you can get many marks for a small amount of writing. Don't ramble on and say the same thing several times over, or wander off into answers that have nothing to do with the question. As a general rule, there will be twice as many answer lines as marks. So you should try to answer a 3-mark question in no more than six lines of writing. If you are writing much more than that, you almost certainly haven't focused your answer tightly enough.

Look carefully at exactly what each question wants you to do. For example, if it asks you to 'explain', then you need to say how or why something happens, not just what happens. Many students lose large numbers of marks by not reading the question carefully.

Take care with multiple-choice questions. It is easy just to see an answer that you think is correct and choose that one, without looking carefully at the others. There are always four choices — A, B, C and D. There will be boxes beside each one, and you tick the correct box. (Note that this format is not used in the multiple-choice questions in this student guide.) If you change your mind, it is important to cross out your 'old' answer clearly and indicate your new one.

Comments

Each question is followed by a brief analysis of what to watch out for when answering the question (shown by the icon **e**). All student responses are then followed by comments that indicate where credit is due. These are preceded by the icon **e**. In the weaker answers, the comments also point out areas for improvement, specific problems, and common errors such as lack of clarity, weak or non-existent development, irrelevance, misinterpretation of the question and mistaken meanings of terms.

■ Sample paper 1

Question 1

Stem rust fungus, *Puccinia graminis*, is a serious pest of wheat. A new race of stem rust fungus, Ug99, was first detected in Uganda in 1988. This race, which has severe effects on wheat production, has now spread to many other wheat-growing countries. Ug99 poses a severe threat to wheat production worldwide.

(a) Describe how stem rust fungus is transmitted from one wheat plant to another. (2 marks)

(b) Outline the pathogenic effects of stem rust on wheat. (3 marks)

(c) Many wheat varieties contain the gene *Sr31*, which has conferred resistance to stem rust in the past, but does not confer resistance to Ug99.
Suggest how natural selection could have resulted in the emergence of Ug99. (4 marks)

(d) Genes *Sr22* and *Sr25*, found in a few varieties of wheat, do confer resistance to Ug99. However, these wheat varieties are not widely grown, and are not adapted for producing high yields in most countries.
Suggest how these genes could be incorporated into other wheat varieties, *without* using gene technology. (2 marks)

Total: 11 marks

e Parts (a) and (b) of this question are a straightforward test of your knowledge of how stem rust fungus is transmitted, and its pathogenic effects. To answer part (c) you will have to think back to your earlier work on natural selection.

> **Student A**
>
> **(a)** The spores ✓ are spread by the wind ✓.

e **2/2 marks awarded** This is just enough for both marks.

> **(b)** The fungus makes red rusty patches on the wheat. It feeds on the wheat ✓, so the wheat doesn't grow very well and its yields are much lower than usual.

e **1/3 marks awarded** Everything that the student says in this answer is correct but it is lacking in good detail. Much more than this is expected in an answer at A-level.

> **(c)** The fungus might have mutated to become resistant ✗ to the gene. The resistant fungi would be more likely to survive and reproduce ✓, passing on their mutations to their offspring ✓. The mutation could have been an insertion, deletion or substitution. Mutations often happen during meiosis.

ⓔ 2/4 marks awarded Poor choice of words has meant that the student has not been given a mark for the idea of a mutation happening in the fungus, because it sounds as though the fungus has done this intentionally. The second sentence of the answer is much better, and gets 2 marks. The last two sentences do not answer the question.

(d) Breed together a wheat plant that has *sr22* and *sr25* with a different wheat plant that has high yields ✓. Select those offspring that are resistant to Ug99 and also have high yields ✓. Breed them together for many generations, choosing the best ones each time ✓.

ⓔ 2/2 marks awarded This is a good answer, giving a clear outline of a suitable approach to a breeding programme.

Student B

(a) Spores ✓ are produced by the fungus growing on an infected plant. The spores are carried on the wind ✓ over long distances.

ⓔ 2/2 marks awarded Student B has made two clear points — that the fungus spreads as spores, and that these are carried on the wind.

(b) The fungus produces hyphae that penetrate the wheat tissues, usually growing through a stoma. The hyphae then grow through cell walls into the cells, taking nutrients ✓ from them. They also block the spaces between the cells, so the cells do not get fresh supplies of nutrients from other cells or from phloem tissue ✓. So the cells die ✓. The fungus does not kill the plant, but it greatly reduces its growth ✓ and the production of grains (seeds) ✓.

ⓔ 3/3 marks awarded This answer is probably a little too long for a 3-mark question, but it is a well-focused response that relates closely to the question and gives several clearly expressed and relevant points.

(c) A random mutation ✓ in the fungus might have caused a change in a gene, perhaps making a protein ✓ that helped to make it resistant to the product ✓ of the wheat gene *Sr31*. The individual fungus with this mutation could infect wheat plants that other fungi could not ✓. So this fungus was able to grow and reproduce ✓ and spread to other wheat plants, without any competition from all the other forms of the fungus. The gene giving resistance would be transferred to all of its offspring ✓, producing a whole fungus population with resistance.

ⓔ 4/4 marks awarded This is an excellent answer, working step by step through the probable process in the sequence in which it might have happened. There is good use of terminology, and the student is careful to make clear when the answer is referring to the gene in the fungus, and when it refers to the gene in the wheat. Six good points have been made, so the maximum 4 marks can be awarded.

(d) Wheat plants that are known to have these resistance genes could be bred with others that don't have them but are well adapted for growing in particular conditions ✓. Some of their offspring would inherit these genes.

ⓔ 1/2 marks awarded The answer begins well, but more is needed for a second mark. Student B could have explained that the offspring would be tested for resistance, by exposing them to Ug99 spores; only those that did not become infected, and that also had other useful characteristics such as producing high yields, would then be used for breeding in the next generation.

Question 2

The thyroid hormone T4 can act as a transcription factor. An investigation was carried out into the effects of T4 on the expression of three genes in the cornea of developing chick embryos.

Corneas of 9-day-old chick embryos were either left untreated or injected with 2.5 µg of T4. The quantities of mRNA transcribed from four different genes, *DIO3*, *THRB*, *OGN* and *CHST1*, were measured over the next 4 days. These quantities were compared against the quantity of mRNA transcribed from a gene called *GAPD*, which is expressed uniformly throughout this period of chick development and is known not to be affected by T4.

The diagram shows the results.

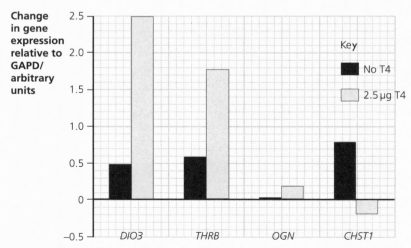

(a) Suggest why the expression of the genes was compared against that of *GAPD*, rather than being recorded as a straightforward quantity of mRNA produced. (2 marks)

(b) (i) Describe the effects of T4 on the expression of the four genes. (3 marks)

(ii) Explain how T4 could have these effects. (3 marks)

Total: 8 marks

e You will probably need to read the information several times, and look carefully at the graph, before you will fully understand how the experiment was conducted, and what the results were. It is essential to give yourself time to absorb this, because all of the questions depend on it.

Student A

(a) So that it was a fair test.

---

e **0/2 marks awarded** This is not enough for a mark.

(b) (i) T4 increases the expression of *DIO3*, *THRB* and *OGN* ✓ and decreases it for *CHST1* ✓.

---

e **2/3 marks awarded** This is correct as far as it goes but more is needed for a third mark.

(b) (ii) It could be a transcription factor ✓. It could bind to the DNA where a gene starts ✓ and stop enzymes binding with it so it won't be used.

---

e **2/3 marks awarded** More detail should be provided about the effect of the transcription factor on the gene for full marks.

Student B

(a) This is in case all the genes are expressed more, and more mRNA is made, as the chick embryo gets older and gets bigger ✓. We know GAPD is expressed just the same all the time, so this gives us a sort of fixed point ✓ that we can compare the other genes against.

---

e **2/2 marks awarded** This is a good answer.

(b) (i) T4 increases the expression of *DIO3*, *THRB* and *OGN* ✓. For *DIO3* it goes up from just below 0.5 to 2.5, and for *THRB* it goes from just above 0.5 to 1.8, so the effect is larger for *DIO3* ✓. T4 decreases the expression of *CGST1* ✓.

---

e **3/3 marks awarded** Three clear and relevant points are made.

(b) (ii) Some hormones act as transcription factors ✓, or sometimes they activate transcription factors in the cell. The T4 could go into the cell through the cell membrane and into the nucleus and bind with the DNA ✓ at a particular point. This could stop the enzymes binding with it ✓ that are needed for transcription ✓, so the gene would not be transcribed into mRNA and would not be expressed. Or it might make it easier for the enzymes to bind ✓, so it would be expressed more than usual.

ℯ **3/3 marks awarded** This is a clear and full answer.

Question 3

The flow diagram shows part of the metabolic pathway of glycolysis.

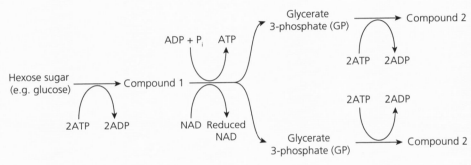

(a) (i) What is compound 1?

  A   fructose

  B   hexose phosphate

  C   lactate

  D   pyruvate   (1 mark)

  (ii) What is compound 2?

  A   ethanol

  B   hexose phosphate

  C   lactate

  D   pyruvate   (1 mark)

(b) In which part of the cell does this metabolic pathway take place?

  A   chloroplast

  B   cytoplasm

  C   Golgi body

  D   mitochondrion   (1 mark)

(c) (i) Describe how compound 2 is converted to lactate in a human muscle cell, if oxygen is not available in the cell.   (2 marks)

  (ii) Describe what happens to the lactate produced.   (3 marks)

Total: 8 marks

ℯ This question is a straightforward test of your knowledge of glycolysis.

Student A

(a) (i) A ✗

   (ii) D ✓

e **1/2 marks awarded** Compound 1 should be B, hexose phosphate.

(b) B ✓

e **1/1 mark awarded**

(c) (i) This is anaerobic respiration. The pyruvate is changed into lactate so it doesn't stop glycolysis happening.

e **0/2 marks awarded** Student A has not answered the question.

(c) (ii) It goes to the liver, which breaks it down ✓. This needs oxygen, which is why you breathe faster than usual when you've done a lot of exercise.

e **1/3 marks awarded** Again, much more is needed for full marks. The comment about breathing rate is correct but is not relevant to this particular question.

Student B

(a) (i) B ✓

   (ii) D ✓

e **2/2 marks awarded**

(b) B ✓

e **1/1 mark awarded**

(c) (i) It is combined with reduced NAD ✓, which is oxidised back to ordinary NAD. The enzyme lactate dehydrogenase ✓ makes this happen.

e **2/2 marks awarded**

(c) (ii) The lactate diffuses into the blood and is carried to the liver cells ✓. They turn it back into pyruvate again ✓, so if there is oxygen it can go into a mitochondrion and go through the Krebs cycle. Or the liver can turn it into glucose again ✓, and maybe store it as glycogen.

e **3/3 marks awarded** All the points made are correct and relevant.

Question 4

The diagram shows the structure of a chloroplast.

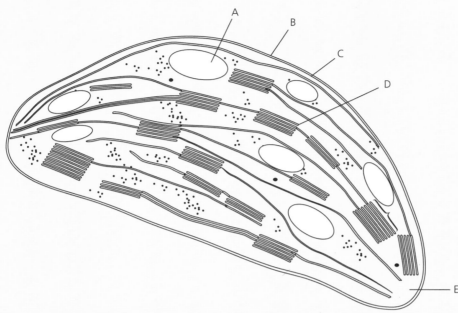

(a) Give the letter of the part of the chloroplast where each of the following takes place.

   (i)   fixation of carbon dioxide   (1 mark)

   (ii) the light-dependent reactions   (1 mark)

(b) A grass adapted for growing in a tropical climate was exposed to low temperatures for several days. The membranes of part D moved closer together, so that there was no longer any space between them. This prevented photophosphorylation taking place.

   Explain how this would prevent the plant from synthesising carbohydrates.   (4 marks)

(c) Two groups of seedlings were grown in identical conditions for 2 weeks. One group was then grown in high-intensity light and the other group in low-intensity light, for 4 weeks.

   Each group of plants was then placed in containers in which carbon dioxide concentration was not a limiting factor. They were exposed to light of varying intensities and their rate of carbon dioxide uptake was measured.

   The results are shown in the graph.

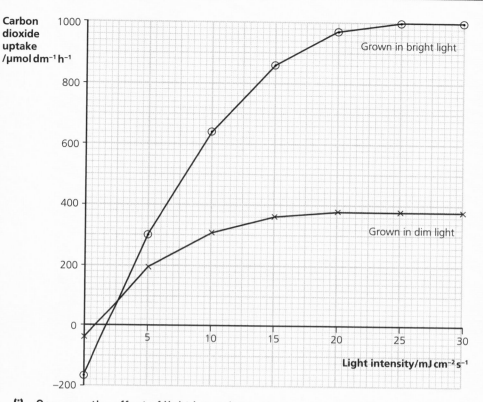

(i) Compare the effect of light intensity on the two groups of plants between 0 and 30 mJ cm^{-2} s^{-1}.

(4 marks)

(ii) Suggest why the plants gave out carbon dioxide at very low light intensities.

(2 marks)

(iii) Suggest two differences in the two groups of plants that could have been caused by their exposure to different intensities of light as they were growing, and that could help to explain the results shown in the graph.

(2 marks)

Total: 14 marks

ⓔ Part (a) is straightforward, but (b) and (c) will require careful thought. Read all the information at least twice before you try to think out good answers to the questions.

Student A

(a) (i) E ✓

(ii) D ✓

ⓔ **2/2 marks awarded**

(b) It would not be able to make any ATP ✓, which is needed for the Calvin cycle ✓. Without the Calvin cycle it would not be able to make carbohydrates.

ⓔ **2/4 marks awarded** This is correct but lacking in detail.

> **(c) (i)** Neither of the groups took up any carbon dioxide when there was no light ✓. Then the quantity of carbon dioxide increased dramatically for the bright light plants, and slowly for the dim light plants. Then it levelled out, lower for the dim light plants than for the bright light ones ✓. The dim light plants levelled out at 380 and the bright light ones at $1000\,\mu mol\,dm^{-2}\,h^{-1}$ ✓.

ⓔ **3/4 marks awarded** Some good comparative points are made here. However, a fundamental error is that the answer is expressed as though the x axis showed time — for example, using the word 'slowly' and 'then'. It is also not a good idea to use terms such as 'dramatically'. See student B's answer for a better way of expressing these points.

> **(c) (ii)** They could not photosynthesise ✓, so the carbon dioxide in their leaves just went back out into the air again.

ⓔ **1/2 marks awarded** One correct point is made here.

> **(c) (iii)** The ones that had grown in the bright light could have bigger leaves and more chlorophyll ✓.

ⓔ **1/2 marks awarded** The suggestion about bigger leaves is not correct. Even if the plants did have bigger leaves, this would not affect the results, because the carbon dioxide uptake is measured per unit area (look at the units on the y axis of the graph). The second point is a good suggestion.

Student B
> **(a) (i)** E ✓
>
> **(ii)** D ✓

ⓔ **2/2 marks awarded**

> **(b)** No ATP would be made ✓ in the light-dependent reaction, so there would not be any available for the light-independent reactions, where carbohydrates (triose phosphate) are made in the Calvin cycle ✓. ATP is needed to convert GP to triose phosphate ✓ (along with reduced NADP) and also to help regenerate RuBP ✓ from the triose phosphate so the cycle can continue.

ⓔ **4/4 marks awarded** This is all correct and with good detail.

(c) (i) Below about 0.5 light intensity, both groups gave out carbon dioxide ✓. As light intensity increased, the amount of carbon dioxide taken up by the plants grown in bright light increased more steeply ✓ than for the ones grown in dim light. In the group grown in dim light, the maximum rate of carbon dioxide uptake was $380\,\mu mol\,dm^{-2}\,h^{-1}$, whereas for the ones in bright light it was much higher ✓, at $1000\,\mu mol\,dm^{-2}\,h^{-1}$ ✓. For the bright light plants the maximum rate of photosynthesis was not reached until the light intensity was $25\,mJ\,cm^{-2}\,s^{-1}$, but for the ones grown in dim light the maximum rate was reached at a lower ✓ light intensity of $20\,mJ\,cm^{-2}\,s^{-1}$.

🅔 **4/4 marks awarded** This is a good answer with some comparative figures quoted (with their units). Note the avoidance of any vocabulary that could be associated with time.

(c) (ii) When the light intensity was very low, the plants would not be able to photosynthesise so they would not take up any carbon dioxide ✓. However, they would still be respiring (they respire all the time) so their leaf cells would be producing carbon dioxide ✓, which would diffuse out into the air. Normally, this carbon dioxide would be taken up by the cells for photosynthesis.

🅔 **2/2 marks awarded** This is a good answer.

(c) (iii) The plants grown in the light would probably be a darker green because they would have more chlorophyll ✓ in their chloroplasts, so they would be able to absorb more light and photosynthesise faster. They might also have more chloroplasts in each palisade cell ✓. And leaves sometimes produce an extra layer of palisade cells if they are in bright light ✓.

🅔 **2/2 marks awarded** This answer actually contains three points, and two of them have been explained, which was not required. Student B could have got 2 marks with a much shorter answer. Nevertheless, this answer shows good understanding of the underlying biology.

Question 5

The polymerase chain reaction, PCR, is widely used in many branches of gene technology.

(a) Describe the purpose of using the polymerase chain reaction. (2 marks)

(b) During the polymerase chain reaction, the temperature of the reacting mixture is repeatedly changed, allowing different enzymes to work.

    (i) Outline the main stages of the polymerase chain reaction. (6 marks)

    (ii) Explain why enzymes work best at a particular temperature. (5 marks)

Total: 13 marks

Questions & Answers

ⓔ Part (b) (i) asks you to 'outline', so you are not expected to write about each stage in detail — just enough to make clear that you know the sequence of steps and can summarise each one.

Student A

(a) To amplify the DNA ✓.

ⓔ **1/2 marks awarded** This is correct but not enough for 2 marks.

(b) (i) The temperature starts off very hot, then it is cooled and then it is heated up again.

ⓔ **0/6 marks awarded** There is not enough here for any marks.

(b) (ii) They work faster when it gets warmer, then when it goes above their optimum temperature they get denatured ✓ and stop working.

ⓔ **1/5 marks awarded** This is a description (and not a very good one), not an explanation — except for the mention of denaturation.

Student B

(a) If you have only a tiny amount ✓ of DNA you can make millions of copies ✓ of it, so you have enough to do DNA profiling ✓.

ⓔ **2/2 marks awarded** This is brief but absolutely correct.

(b) (i) First the DNA is split apart into two strands ✓ by heating it to 95°C ✓. Then primers are attached ✓ at a temperature of 65°C ✓. Then DNA polymerase ✓ starts at the primers and makes a copy of each strand of the DNA ✓, at a temperature of 72°C ✓. Then the new DNA molecules are heated up and split apart again, and so on.

ⓔ **6/6 marks awarded** A short answer that is full of correct detail.

(b) (ii) Enzymes don't work very fast at low temperatures because they don't have much kinetic energy ✓ so they don't bump into their substrate very often ✓. But when it is too hot the hydrogen bonds ✓ in the enzyme molecule break so the molecule loses its shape ✓ and the substrate doesn't fit in the active site ✓. In between there is the perfect temperature for reacting.

ⓔ **5/5 marks awarded** Another short answer that is packed with relevant facts.

Question 6

The diagram below shows part of the small intestine of a mouse. The inner wall of the small intestine is made up of many tiny folds called villi. Between the villi are indentations known as crypts of Lieberkühn. The cells on the villus surface are specialised for the production of digestive enzymes and the absorption of the products of digestion.

Cells at the upper surfaces of the villi are constantly shed. They are replaced by new cells which are produced from stem cells in the crypts and steadily work their way upwards. As they move upwards they differentiate.

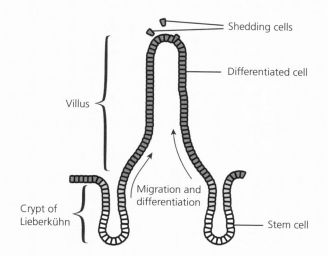

(a) Give the term for a group of similar cells such as those covering the surface of a villus.
(1 mark)

(b) Explain what is meant by the term stem cell.
(2 marks)

(c) State whether or not the stem cells in the crypts of Lieberkühn are totipotent. Explain your answer.
(2 marks)

(d) Suggest the changes that will take place in the cells as they differentiate.
(4 marks)

(e) (i) Outline the reasons why embryonic stem cells may be useful in medical therapies.
(4 marks)

  (ii) Suggest why some people may have ethical objections to the use of embryonic stem cells for such therapies.
(2 marks)

Total: 15 marks

🄴 Note that parts (d) and (e) (ii) ask you to 'suggest...'. You should use your knowledge to provide biologically sound answers, thinking about how your understanding of differentiation and stem cell therapies can be applied in these particular situations.

Questions & Answers

(a) Tissue ✓

e **1/1 mark awarded**

(b) A cell that can divide ✓ and produce new cells.

e **1/2 marks awarded** This is partly correct but not enough for the second mark.

(c) No, because only embryonic stem cells are totipotent.

e **0/2 marks awarded** Although 'no' is correct, there is not a mark for this; both marks are for the explanation. This is not a good answer, because it does not relate directly to the cells in the question; nor does it explain what totipotent means.

(d) They will become able to produce digestive enzymes ✓ and will move up the villus.

e **1/4 marks awarded** Much more detail is expected than is given in this answer. Student A has not described what will change in the cells as they differentiate.

(e) (i) They could be used to cure Alzheimer's or Parkinson's ✓. You could put stem cells into the brain and they will make new cells that work properly.

e **1/4 marks awarded** Student A gives two examples of diseases that could be treated with stem cells but does not mention why embryonic stem cells would be especially useful, or how the stem cells might treat the disease.

(e) (ii) To get the cells, you have to kill embryos ✓.

e **1/2 marks awarded** Once again, this answer does not have sufficient depth.

(a) Tissue ✓

e **1/1 mark awarded**

(b) It is a cell that has not differentiated ✓, and can keep on dividing to produce new cells ✓. These new cells can then differentiate into specialised cells.

e **2/2 marks awarded** This is a good answer.

(c) No. Totipotent stem cells are able to produce all the different kinds of specialised cells ✓. These stem cells only make cells for the surface of the villus ✓.

ⓔ 2/2 marks awarded This is a clear and entirely correct answer. Student B has explained what 'totipotent' means, and then applied this knowledge to the particular stem cells in the question.

(d) Particular sets of genes would be switched on or off ✓. The cells would get more endoplasmic reticulum ✓ and Golgi apparatus ✓.

ⓔ 3/4 marks awarded This is a good answer with some specific information. Student B has only thought about enzyme synthesis, however, and not about the other function of the cells, which is absorption. The cells would also increase the number of mitochondria they contain (to provide energy for protein synthesis and active transport) and will develop microvilli (lots of small folds on their surfaces), to increase the surface area for absorption.

(e) (i) Some diseases, like Parkinson's ✓, are caused by certain groups of cells not working ✓. In Parkinson's disease, cells in the brain that should produce dopamine stop doing this ✓. If you could put stem cells into the brain then they might be able to divide and make new dopamine-producing cells ✓. This would mean the person might not need to keep taking drugs. Embryonic stem cells would be especially useful because they are able to form all the different kinds of cell ✓ in the body.

ⓔ 4/4 marks awarded This is an excellent answer that gets the maximum marks available.

(e) (ii) You have to kill embryos to get the cells ✓.

ⓔ 1/2 marks awarded This is correct as far as it goes but more is needed for the second mark. For example, Student B could have explained that the embryos are likely to be ones that have been produced by in vitro fertilisation and that could otherwise be frozen and stored for later use. Another mark-worthy comment would be that people might want to produce embryos just so that they could get stem cells from them.

Question 7

IgA is an immunoglobulin (antibody) that is particularly important in controlling infections in the tissues lining the respiratory passages. The graph shows how the quantity of IgA in the saliva of an elite kayaker changed during a 13-day intensive training period.

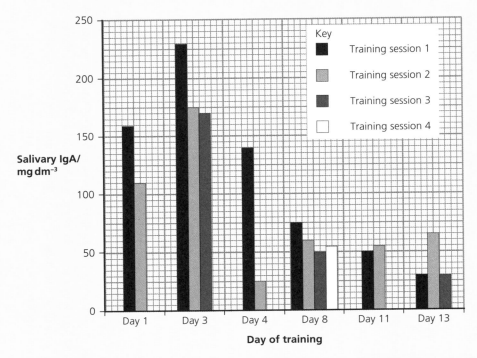

Day of training

(a) (i) Calculate the percentage change in salivary IgA concentration between the first and second training sessions on day 1. Show your working. (3 marks)

(ii) Describe the changes in salivary IgA during the 13-day training period. (4 marks)

(b) It has been known for many years that intensive training is often associated with an increased likelihood of suffering from an upper respiratory tract infection (for example, a cold). Discuss the extent to which the data in the graph provide evidence for a causal link between training and upper respiratory tract infections. (3 marks)

Total: 10 marks

ⓔ Remember to show all of your working really clearly for (a) (i). Note that (a) (ii) asks you to *describe* the changes, not *explain* them.

Student A

(a) (i) $160 - 110 = 50$ ✓

$\dfrac{50}{160} \times 100 = 31\%$ ✓

ⓔ **2/3 marks awarded** The calculation is correct but the answer does not state whether the change is positive or negative.

(a) (ii) The concentration went down overall ✓. The highest amount was after the first session on day 1, and the lowest was after the first session on day 13. However, on day 2 it went up ✓ for some reason, and was higher after all the training sessions than it was at any point on day 1. On days 1, 3, 4 and 8 it was lower after the second training session than after the first training session ✓, but on days 11 and 13 the opposite happened ✓.

🅔 **4/4 marks awarded** Student A has picked up the most significant trends on the graph.

(b) IgA helps to fight off viruses that might give you a cold, so it makes sense that if there is less IgA then you stand more chance of getting a cold ✓. So this does suggest there is a link between getting a cold and training.

🅔 **1/3 marks awarded** Student A has not fully focused on the issue of whether this particular set of data indicates a causal link between training and getting a cold.

Student B

(a) (i) Difference between first and second sessions = 160 – 110 = 50 ✓ $mg\,dm^{-3}$

So percentage change = $\dfrac{50}{160} \times 100$ ✓ = –31% ✓

🅔 **3/3 marks awarded** This is entirely correct, including showing that the percentage change is a decrease.

(a) (ii) The concentration went down after the second training session on each of the first 8 days for which we have information ✓, but after that it went up. It also went up between day 1 and day 2 ✓, but after that it kept going down each day ✓. The biggest change between training sessions was on day 4 ✓, when it went down about 80% ✓ after the second training session.

🅔 **4/4 marks awarded** The main trends and patterns have been picked out, and a calculation has been made relating to one of the significant changes.

(b) The graph does not tell us anything about infections, only about the levels of IgA ✓. So the data could support the hypothesis that training causes more respiratory infections, because it would make sense that if there is less IgA then you have a bigger chance of getting infections ✓. But the data don't actually prove this, so we would have to measure these same things again in lots of different people and also measure the number of colds they got ✓. Even then we wouldn't be absolutely sure there was a causal link, because there might be some other factor we haven't measured that is causing the difference ✓.

@ **3/3 marks awarded** Some good points have been made here — the first one is especially important.

Question 8

The diagram shows a vertical section through a human heart.

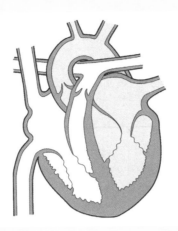

(a) On the diagram, label:

    (i) the position of the sinoatrial node (SAN)   (1 mark)

    (ii) the position of the atrioventricular node (AVN)   (1 mark)

    (iii) the path taken by the wave of electrical activity during one heart beat.   (2 marks)

(b) Heart muscle tends to use fatty acids as the main respiratory substrate, rather than glucose. Fatty acids are first converted to acetyl coenzyme A before taking part in the metabolic pathways of respiration.

    (i) Explain how this information helps to explain why heart muscle is unable to respire anaerobically to any great extent.   (3 marks)

    (ii) Outline how reduced NAD is used to produce ATP in the inner mitochondrial membranes of heart muscle.   (4 marks)

Total: 11 marks

@ If you have learnt this part of your work well, this should be a straightforward question. Part (b) (i) is the most difficult, as you will need to link together the information provided here and your knowledge of anaerobic respiration and the Krebs cycle.

Student A

(a)

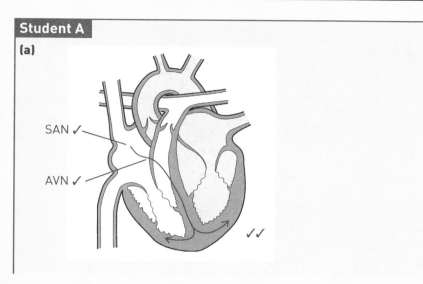

SAN ✓

AVN ✓

✓✓

ⓔ **4/4 marks awarded**

(b) (i)   It couldn't do glycolysis if it doesn't use glucose ✓.

ⓔ **1/3 marks awarded** There is the beginning of a correct answer here, but much more needs to be included in the explanation.

(b) (ii) The reduced NAD gives its electrons ✓ to the first thing in the electron transport chain ✓. The electron goes along the chain and is used to make ATP. This happens inside the mitochondria and it is called oxidative photophosphorylation ✗.

ⓔ **2/4 marks awarded** The answer begins correctly but we are not told any detail about how the ATP is made. The term 'photophosphorylation' is incorrect.

Student B

(a)

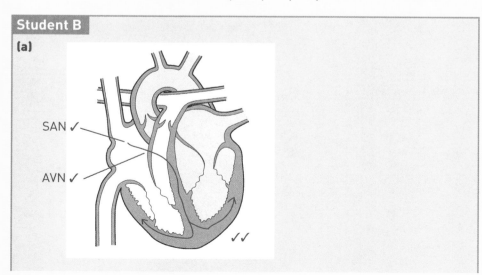

SAN ✓

AVN ✓

✓✓

e 4/4 marks awarded

> **(b) (i)** The fatty acids will have to go into the mitochondria and when they turn into acetyl coenzyme A they will go through Krebs cycle ✓. This can only happen when there is oxygen available ✓. Anaerobic respiration can only be done using glycolysis ✓ as a starting point, which uses glucose and changes it to pyruvate, which is then converted to lactate ✓. Fatty acids can't be used for glycolysis.

e 3/3 marks awarded This is a good answer.

> **(b) (ii)** Hydrogens from the reduced NAD are released ✓ and split into protons and electrons. The electrons go along the carriers ✓ in the inner mitochondrial membrane, losing energy ✓ as they are passed along. The energy is used to pump hydrogen ions into the space between the membranes ✓. When the hydrogens diffuse back out, they go through ATP synthases ✓, which use the energy to combine ADP and Pi to make ATP. This is called chemiosmosis ✓.

e 4/4 marks awarded This is entirely correct.

■ Sample paper 2

Question 1

A student extracted pigments from a sample of leaves and then made a chromatogram to identify the pigments.

(a) Describe how the student could extract the pigments from the leaves, ready to use them for chromatography.　　　　　　　　　　　(3 marks)

(b) Describe how the student should set up and run the chromatography experiment.　　　　　　　　　　　(6 marks)

(c) The diagram shows the chromatogram that the student obtained.

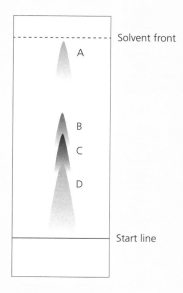

(i) Using a ruler marked in mm to make your measurements, calculate the *Rf* value for spot C. Show your working.　　　　　　(2 marks)

(ii) Calculate the percentage error for the *Rf* value that you have calculated. Show your working.　　　　　　　　　(3 marks)

Total: 14 marks

ⓔ This is a procedure that you should have carried out during your A-level course, so use your experience to give clear and full descriptions in your answers to (a) and (b). For part (c), show each step in your calculations clearly.

Student A

(a) Grind up the leaves with a solvent ✓. Grinding them will open up the cells, so the pigment can get out and dissolve in the solvent.

ⓔ 1/3 marks awarded The first sentence gains 1 mark. However, there is no detail — for example, what solvent is used, that it needs to be kept cold, or how the leaves will be ground. Student A also fails to describe how the extract will be prepared ready for chromatography, which requires a clear solution with no leaf pieces in it. The second sentence does not answer the question, as it attempts to *explain why* the process is carried out, rather than just *describing* it.

(b) Draw a line on some chromatography paper, and put spots of the pigment onto it ✓. Then stand the paper in a jar with some solvent and leave it. The solvent will soak up the paper and take the pigments with it, leaving different pigments at different places on the paper. Then you can measure how far each spot has gone from the start line and where the solvent front has got to, and then use these to calculate the *Rf* value.

ⓔ 1/6 marks awarded Student A has written a lot of material that does not answer the question. Only the first two sentences are relevant to the question, which asks about what the student should do to set up and run the chromatogram. There is not enough detail in those first two sentences to gain more than 1 mark.

(c) (i) $\dfrac{113}{56}$ ✓ = 2.02

ⓔ 1/2 marks awarded Student A has not fully explained what the working shows. The two figures — 56 and 113— are correct measurements, in mm, for the distance of the pigment spot and the solvent front from the start line, but it would have been much better if the answer made this clear. A rather generous mark has been given for these two figures. The calculation is incorrect. To calculate *Rf*, the distance moved by the pigment is divided by the distance to the solvent front, so this calculation is 'upside-down'.

(c) (ii) Each measurement could be out by 0.5 mm ✓, so overall it could be out by ±1.0 mm.

ⓔ 1/3 marks awarded It is correct to say that each measurement could be out by 0.5 mm, and a mark has been given for this. However, the question asks for the percentage uncertainty to be calculated, and there is no attempt to do this. It is not appropriate just to add the two values together in this case.

Student B

(a) Cut the leaves into small pieces. Grind them in a pestle and mortar ✓ with a little cold ✓ solvent. Then pour off the liquid, leaving any pieces of leaves behind ✓.

e **3/3 marks awarded** Student B has written a concise and correct answer, including the detail that the solvent should be cold. The answer would be even better if they had stated what the solvent could be. It would also be better to pour the extract through muslin to make sure that there are no pieces of leaf remaining in it. Nevertheless, this gets full marks.

> **(b)** Take a piece of chromatography paper. Try not to touch it with your fingers, as you can leave greasy marks that might affect how the pigments and solvent move up the paper ✓. Use a pencil and ruler to draw a line about 2 cm from one end of the paper ✓. Then get a glass pipette with a really narrow point and dip it into the extract. Put a spot of the extract exactly onto the pencil line. Keep doing this, trying to make a really tiny spot but with a lot of extract on it ✓.
>
> Next pour some solvent into a gas jar, to a depth less than 2 cm ✓. Stand the chromatography paper in the solvent — the solvent must not reach the pencil line. Put a lid over the jar ✓ and leave the chromatogram to run. When the solvent has got nearly to the top of the paper, take the paper out ✓. Lie it on a clean surface and draw a line where the solvent has got to ✓. Then draw round all the spots ✓, because if you don't you won't be able to see where they were when they have faded.

e **6/6 marks awarded** It is clear that Student B has used this technique because they have given such a clear and thorough description. The answer is written in an entirely logical sequence and is easy to follow.

> **(c) (i)** distance of solvent front from start line = 113 mm
> distance of top of spot C from start line = 56 mm ✓
> Therefore $Rf = \dfrac{56}{113} = 0.50$ ✓

e **2/2 marks awarded** Both measurements are correct, and the Rf value has been correctly calculated. The working is clearly shown.

> **(c) (ii)** The ruler was marked in mm, so the uncertainty in each measurement is ±0.5 mm ✓.
>
> percentage uncertainty in first measurement $= \dfrac{0.5}{113} \times 100 = 0.44\%$
>
> percentage uncertainty in second measurement $= \dfrac{0.5}{56} \times 100 = 0.89\%$ ✓
>
> The two numbers were then divided, so the total uncertainty is 0.44 + 0.89 = 1.33% ✓.

ⓔ **3/3 marks awarded** This is a perfect answer. Student B knows that the uncertainty in a measurement is half the smallest division on the scale. They also know how to calculate percentage uncertainty, and also how to add these percentages together to calculate the overall percentage uncertainty when numbers are divided.

Question 2

The rate of aerobic respiration of yeast cells can be measured using alkaline phenolphthalein indicator. This changes colour from purple to colourless when the pH of a solution falls to a certain level.

A student set up four test tubes. Each tube contained the same volumes of yeast suspension, glucose solution and alkaline phenolphthalein indicator.

Ethanol was also added to three of the tubes, to produce ethanol concentrations as shown in the table. The table also shows the time taken for the indicator in the tube to become colourless.

Tube	Ethanol concentration/%	Time taken to become colourless/s
1	0	32
2	1.5	96
3	12.5	193
4	25.0	Did not become colourless

(a) Explain why the indicator lost its colour in tubes 1, 2 and 3. (3 marks)

(b) (i) With reference to the results in the table, describe the effect of ethanol on the rate of respiration in yeast. (2 marks)

    (ii) Suggest reasons for the effect you described in (i). (2 marks)

(c) State two variables, other than the volumes of yeast suspension, glucose solution and indicator, that should be kept constant during this experiment, and explain how the student could achieve this. (4 marks)

Total: 11 marks

ⓔ This question tests knowledge and understanding of designing an experiment, as well as respiration. Read it carefully, and make sure you understand why the indicator became colourless.

Student A

(a) This is because the yeast made carbon dioxide ✓, which makes an acid when it dissolves in water.

ⓔ **1/3 marks awarded** This is a correct answer but the explanation is not full enough for a 3-mark question.

(b) (i) The ethanol inhibited respiration ✓. At a concentration of 25% it stopped it completely ✓.

ⓔ 2/2 marks awarded A correct general point is made and then a specific one.

(b) (ii) Perhaps the alcohol is toxic ✓ to yeast.

ⓔ 1/2 marks awarded This is a reasonable suggestion but more is needed for a second mark.

(c) Temperature ✓. Put it into a water bath ✓. Concentration of glucose solution ✓. Measure all the glucose from the same solution ✓.

ⓔ 4/4 marks awarded Two relevant variables to be controlled are given and outlines (though very brief) of how they would be controlled.

Student B

(a) As the yeast respired ✓, it produced carbon dioxide ✓, which dissolved in the water and made carbonic acid ✓. This is a weak acid, which would lower the pH ✓. The more carbon dioxide, the lower the pH, until it goes below the point at which alkaline phenolphthalein changes from purple to colourless ✓.

ⓔ 3/3 marks awarded This is a thorough explanation.

(b) (i) The greater the concentration of alcohol, the slower the rate of respiration ✓. It seems that respiration did not happen at all when there was 25% ethanol ✓, but we cannot be sure that it might not have eventually changed colour if it had gone on longer.

ⓔ 2/2 marks awarded This is a good answer.

(b) (ii) Perhaps ethanol inhibits one of the enzymes ✓ involved in respiration, by binding with it at a point other than its active site and changing the shape of its active site ✓. Or it might be a competitive inhibitor, competing for the active site with the enzyme's normal substrate ✓.

ⓔ 2/2 marks awarded This is a good suggestion, expanded to give two alternative ways in which this could happen.

(c) Temperature ✓ — place all the apparatus in a thermostatically controlled water bath at about 30°C ✓. Concentration of yeast ✓ — make up a large quantity of yeast suspension, and stir it thoroughly before measuring out the same volume into each tube ✓.

ⓔ 4/4 marks awarded Two suitable suggestions are given and the methods of controlling them are well explained.

Question 3

The diagram shows the bacterium *Mycobacterium tuberculosis*, which causes tuberculosis (TB).

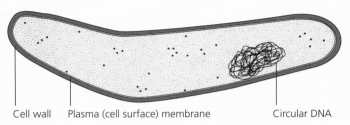

Cell wall   Plasma (cell surface) membrane   Circular DNA

(a) List *three* ways in which the structure of this bacterium differs from that of a virus.   (3 marks)

(b) *M. tuberculosis* is taken up by macrophages, and multiplies inside them.
Explain how this strategy helps to protect *M. tuberculosis* from the immune response by B cells.   (4 marks)

(c) In an experiment to investigate how *M. tuberculosis* avoids destruction by macrophages, bacteria were added to a culture of macrophages obtained from the alveoli of mice. At the same time, a quantity of small glass beads, equivalent in size to the bacteria, were added to the culture.
The experiment was repeated using increasing quantities of bacteria and glass beads.
After 4 hours, the macrophages were sampled to find out how many had taken up either glass beads or bacteria. The results are shown in the graph. The *x* axis shows the initial ratio of bacteria or glass beads to macrophages in the mixture.

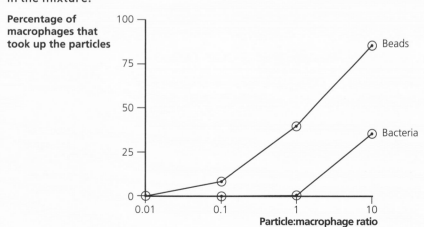

Discuss what these results suggest about the ability of macrophages to take up *M. tuberculosis*.   (3 marks)

(d) When *M. tuberculosis* is present inside a phagosome of a macrophage, it secretes glycolipids that accumulate in lysosomes and prevent them fusing with the phagosome. Explain how this prevents the macrophage from destroying the bacterium.   (3 marks)

(e) With reference to your answer to (d), suggest how natural selection could result in a population of mice that are resistant to *M. tuberculosis*.

(6 marks)

Total: 19 marks

e The most demanding part of this question is (c) because you will have to think carefully about the data in the graph. Part (e) tests your understanding of natural selection, which you will have covered in Topic 3.

Student A

(a) It is made of a cell ✓, it has a cell wall ✓ and it is much larger ✓.

e **3/3 marks awarded** Three correct differences are given.

(b) It stops the B cells seeing them, so they don't make antibodies ✓ against them.

e **1/4 marks awarded** This is not a very clear answer. B cells do not 'see', so this is not a good term to use. The 'they' in the second sentence could refer to either B cells or the bacteria.

(c) The macrophages took up more glass beads than bacteria ✓ so they are not very good at taking up the bacteria ✓.

e **2/3 marks awarded** This is just enough for 2 marks, although the second sentence is weak.

(d) Lysosomes contain digestive enzymes ✓, so if they don't fuse with the phagosome the bacteria won't get digested ✓.

e **2/3 marks awarded** Once again, Student A has the right ideas but does not give enough biological detail to get full marks.

(e) If a mouse had a gene that made a chemical that stopped the bacteria secreting the glycolipids ✓, or if it had a gene that made the lysosomes fuse with the phagosomes even if the lipids were there ✓, that would give it an advantage and it would be more likely to survive ✓.

e **3/6 marks awarded** This answer explains how a mouse could have a selective advantage but it does not go on to explain how a population of resistant mice could arise.

Student B

(a) The bacterium is cellular but a virus is not. ✓ A virus has a capsid made of protein but a bacterium does not ✓. A bacterium has a cell wall, but a virus does not ✓.

ⓔ **3/3 marks awarded** These three points are all correct.

(b) B cells only become active when they meet the specific antigen ✓ to which they are able to respond. If the bacteria are inside a macrophage, then the B cell's receptors won't meet the antigen ✓ on the bacteria. This means that the B cells will not divide to produce plasma cells ✓, and will not secrete antibodies ✓ against the bacteria.

ⓔ **4/4 marks awarded** This is a good answer.

(c) The cells only started to take up any bacteria when the particle:macrophage ratio was 1 ✓. On the other hand, they took up glass beads even when the ratio was only just above 0.01 ✓. When the ratio of particles to macrophages was 10, only about 30% ✓ of the macrophages had taken up bacteria, whereas over 75% of them had taken up glass beads. ✓ This shows the macrophages do take up the bacteria, but not as well as they take up glass beads ✓.

ⓔ **3/3 marks awarded** This is a good answer, which does attempt to 'discuss' by providing statements relating to the relatively low ability of the macrophages to take up the bacteria, but also stating that they do take them up.

(d) Normally lysosomes fuse with phagosomes and release hydrolytic enzymes ✓ into them. These enzymes then hydrolyse (digest) whatever is in the phagosomes ✓. If this doesn't happen, then the bacteria can live inside the phagosomes ✓ without being digested.

ⓔ **3/3 marks awarded** All correct.

(e) Not all mice will have the same alleles ✓, so some mice might happen to have an allele of a gene that means its lysosomes are not sensitive to the glycolipids that the bacteria produce ✓. These mice would be able to destroy the bacteria ✓, so they would not get TB ✓. They would be more likely to survive and reproduce ✓, and pass on their alleles to their offspring ✓. This could happen over several generations, so the advantageous alleles would become more common ✓, until perhaps the whole population had these alleles and was resistant to the bacteria.

ⓔ **6/6 marks awarded** This answer works logically through the mechanism of natural selection, applied to this particular situation. It is clearly expressed.

Question 4

(a) The diagram shows the Calvin cycle.

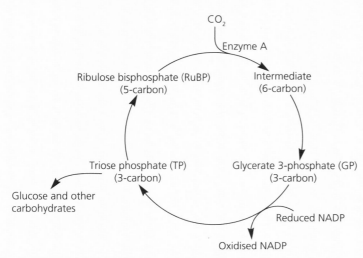

    (i) Name enzyme A. (1 mark)

    (ii) On the diagram, draw and label arrows to show how and where ATP from the light-dependent reaction is used. (2 marks)

    (iii) From the diagram, what is the first carbohydrate made during the Calvin cycle? (1 mark)

(b) An experiment was carried out in which photosynthesising tissues were exposed to light, then to darkness and then to light again. The graph shows the relative amounts of GP (glycerate 3-phosphate) and RuBP (ribulose bisphosphate) during the experiment.

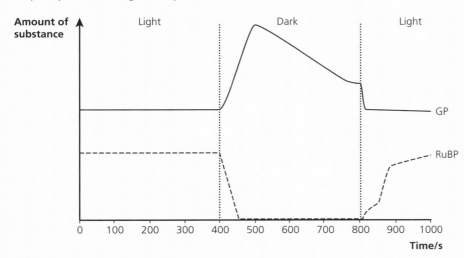

    (i) Describe the effects of light and dark on the amounts of GP and RuBP. (4 marks)

    (ii) Suggest explanations for the changes shown in the graph. (5 marks)

Total: 13 marks

ⓔ The hardest part of this question is (b) (ii). Look for a change in gradient of one of the lines, and think of a logical cause of this change. Then do the same for another change in gradient, until you feel you have made at least five strong points.

Student A

(a) (i) Rubisco ✓

ⓔ **1/1 mark awarded**

(a) (ii)

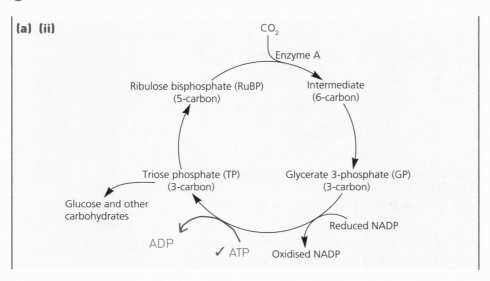

ⓔ **1/2 marks awarded** This one is correct but there should be another one for a second mark.

(a) (iii) Ribulose bisphosphate ✗

ⓔ **0/1 mark awarded** The correct answer is triose phosphate.

(b) (i) When it is light they stay the same. In the dark GP increases but RuBP decreases ✓. When it is light again they go back to normal.

ⓔ **1/4 marks awarded** This poorly worded answer is lacking in detail. 'Stay the same' could mean the same as each other, or constant. 'Go back to normal' does not indicate what 'normal' might be.

(b) (ii) When it is light, the plant can do the light-dependent reactions ✓ and the light-independent reactions, so it can make plenty of RuBP and GP in the Calvin cycle. When it is dark it runs out of reduced NADP and ATP ✓ made in the light-dependent reactions, so it cannot change GP to triose phosphate ✓ and so the GP just builds up ✓.

ⓔ 4/5 marks awarded This is a much better answer than the previous one. Student A works logically through the probable reasons for the patterns shown on the graphs. However, they have not addressed the fall in RuBP during darkness, or the gradual fall in GP after it has risen during darkness.

Student B

(a) (i) RUBISCO ✓

ⓔ 1/1 mark awarded

(a) (ii)

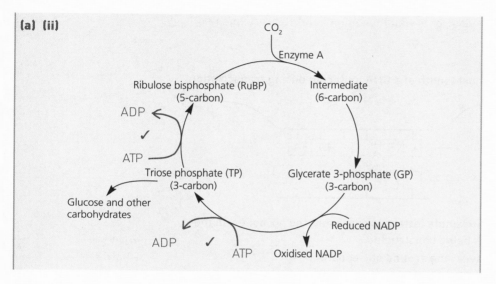

ⓔ 2/2 marks awarded Both correct.

(a) (iii) Triose phosphate ✓

ⓔ 1/1 mark awarded

(b) (i) During the light the amounts of GP and RuBP remain constant ✓. When it is dark, the GP quickly increases to a peak ✓ and then steadily decreases ✓. The RuBP quickly disappears altogether. ✓ When the light comes back on the GP drops quickly ✓ and then goes back to the level it was at the start of the experiment and stays there ✓. The RuBP immediately starts to increase ✓ until it reaches its original level ✓.

ⓔ 4/4 marks awarded This is a thorough description.

(b) (ii) In the dark, there won't be any ATP or reduced NADP ✓ from the light-dependent reaction ✓. So the Calvin cycle gets stuck at GP ✓ because it can't be turned into triose phosphate ✓, so the GP just builds up. As there is no ATP, the triose phosphate (what there is of it, because no more is being made now) can't be changed into RuBP ✓ so this decreases to nothing ✓ as it is all made into GP. When the RuBP all runs out, no more GP is made so it stops increasing ✓. Maybe some of the GP just disintegrates if it can't be made into triose phosphate, which is why it falls.

ⓔ **5/5 marks awarded** This answer is well thought out and clearly explained.

Question 5

(a) The diagram shows a short length of a DNA molecule during transcription.

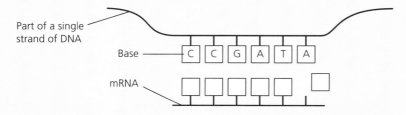

(i) On the diagram, write single letters to represent the six bases on the mRNA strand that is being constructed. (2 marks)

(ii) On the diagram, draw a ring around one codon. (1 mark)

(iii) State the part of an animal cell in which transcription takes place. (1 mark)

(b) Many lengths of DNA (genes) are made up of alternating exons and introns, and these are all used as a template to make an mRNA molecule with a complementary base sequence.

(i) Describe how this mRNA molecule is modified before it is used in the production of a protein molecule. (2 marks)

(ii) Antibodies are immunoglobulins, which are protein molecules. Explain how the post-transcriptional modification of mRNA allows several different forms of an immunoglobulin to be coded for by just one gene. (3 marks)

(iii) Explain why having a very large number of different immunoglobulins is advantageous to a person. (4 marks)

Total: 13 marks

ⓔ This question covers two different parts of the specification — protein synthesis and immunology. You should always be prepared to make links between different areas of biology.

Student A

(a) (i)

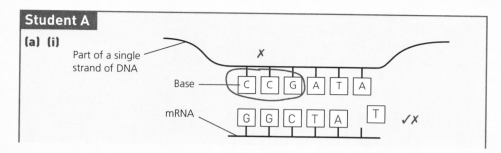

e **1/2 marks awarded** Student A has forgotten that RNA contains U instead of T.

(a) (ii) See diagram

e **0/1 mark awarded** Codons are found on the mRNA molecule, not on DNA.

(a) (iii) Ribosome ✗

e **0/1 mark awarded** Student A has perhaps confused transcription and translation.

(b) (i) The introns are chopped out ✓ and the exons join up with one another ✓.

e **2/2 marks awarded** This is just enough for both marks.

(b) (ii) You can stick the introns together in different ways ✓ so you can get lots of different mRNAs ✓, which can make lots of different sorts of antibody.

e **2/3 marks awarded** The last part of the answer repeats the question, so there is no mark for that.

(b) (iii) You can make an antibody to attack all the different kinds of pathogen that get into the body ✓.

e **1/4 marks awarded** This is correct but nowhere near enough information for a 4-mark question.

Student B

(a) (i)

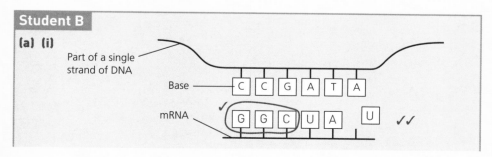

Questions & Answers

e 2/2 marks awarded

> **(a) (ii)** See diagram

e 1/1 mark awarded

> **(a) (iii)** Nucleus ✓

e 1/1 mark awarded

> **(b) (i)** After the pre-mRNA has been made, the introns are separated from the exons ✓. The exons are then linked together to make a continuous chain ✓. This is the mRNA that leaves the nucleus. This is called RNA splicing.

e 2/2 marks awarded This is correct.

> **(b) (ii)** Immunoglobulins are proteins with a similar molecular structure, but with a variable region at one end, which sticks to antigens. One immunoglobulin gene can make many different versions of the protein because the exons can be linked together in different ways ✓, so the base sequence on the mRNA is different ✓ and therefore so is the amino acid sequence ✓ in the protein that is made.

e 3/3 marks awarded This is a clear and correct answer.

> **(b) (iii)** Each immunoglobulin is able to bind with one particular antigen ✓. As each antigen has a different molecular structure, you need many kinds of immunoglobulin to be sure that you have one that can bind with whatever antigen ✓ gets into the body. When the immunoglobulin binds with the antigen, it helps other cells like macrophages to destroy it ✓. So this protects you against infections ✓ and allows you to survive.

e 4/4 marks awarded This is all correct.

Question 6

(a) The diagram shows a sperm cell.

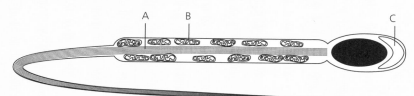

Describe the roles of each of the following parts in the events immediately preceding fertilisation of a female gamete.

(i) part A (1 mark)

(ii) part B (3 marks)

(iii) part C (2 marks)

(b) There are a small number of stem cells in the testes of mice, which are normally able to produce only sperm cells. In 2007 a way was found to 'reprogram' the stem cells so that they become pluripotent.

(i) Explain the meaning of each of the following terms:

stem cell   pluripotent (3 marks)

(ii) The stem cells were 'reprogrammed' by exposing them to various chemicals. Suggest what the 'reprogramming' is likely to cause to happen inside the cell. (2 marks)

(iii) It is hoped that the use of stem cells may eventually become routine medical therapy for several serious diseases. Discuss the reasons why being able to use pluripotent stem cells derived from adult testes may be preferable to using embryonic stem cells for this purpose. (5 marks)

Total: 16 marks

🄮 Note the different mark allocations for the three parts of (a), and construct your answers accordingly. There are 5 marks for (b) (iii), which is often a sign that there is much to be gained by a little planning before you begin your final answer.

Student A

(a) (i) This helps the sperm to swim.

🄮 **0/1 mark awarded** This is true but a bit more information is needed to earn a mark.

(a) (ii) This is a mitochondrion, where respiration happens. This provides energy ✓ for the sperm cell to be able to swim.

🄮 **1/3 marks awarded** This is just enough to get 1 mark. There are 3 marks available, indicating that much more detail is required. The answer needs more specific information about the stages of respiration that take place inside mitochondria, and how 'energy' is provided.

Questions & Answers

(a) (iii) This is the acrosome and it contains enzymes to digest a pathway into the egg ✓.

ⓔ **1/2 marks awarded** There is no credit for naming the acrosome because the question asks about its function, not its name.

(b) (i) A stem cell is a cell that can divide to produce other cells ✓. Pluripotent means it is very strong and able to produce a large number of other cells.

ⓔ **1/3 marks awarded** The definition of a stem cell gives only just enough information to get 1 mark. The definition of pluripotent is not correct.

(b) (ii) The organelles inside the cell would change.

ⓔ **0/2 marks awarded** This may be true but it is not enough for a mark.

(b) (iii) It would mean no embryos will have to be killed to get the stem cells ✓. It will be much easier and cheaper to get stem cells from a man's testes.

ⓔ **1/5 marks awarded** The first sentence makes a good point. The second needs to be developed a little more before it gets a mark — for example, by explaining why it would be 'easier' or 'cheaper'.

Student B

(a) (i) This is a microtubule and they provide the movement ✓ that makes the tail lash from side to side.

ⓔ **1/1 mark awarded** This is correct.

(a) (ii) This is a mitochondrion. Aerobic respiration ✓ happens here, involving the Krebs cycle and oxidative phosphorylation ✓, which makes ATP ✓. This is the sperm cell's source of energy for swimming.

ⓔ **3/3 marks awarded** Three clear and relevant points are made, using terminology well.

(a) (iii) This is the acrosome. When the sperm cell meets an egg the acrosome bursts open and lets out its hydrolytic ✓ enzymes ✓. These help to break down the zona pallisade ✗ around the egg so the sperm nucleus can get in and fertilise it.

ⓔ 2/2 marks awarded There is a mark for knowing that this organelle contains enzymes, and another for the fact that these enzymes are hydrolytic. The reference to digesting the zona pellucida would also have been worth a mark, but the spelling is wrong (it looks like 'palisade', as in 'palisade cell'). Usually spelling does not matter, but when it means that the word could be confused with something else then the student won't be credited. In this case it does not matter, because the answer has already scored the maximum number of marks.

> **(b) (i)** A stem cell is an undifferentiated ✓ cell that can divide to produce new cells that can become specialised ✓. A pluripotent stem cell is one that can produce many different kinds ✓ of specialised cell, but is not capable of producing a whole organism.

ⓔ 3/3 marks awarded This is explained clearly.

> **(b) (ii)** Reprogramming switches on different sets of genes ✓. When a cell gets specialised, certain sets of genes are switched on and others are switched off. Some types of cell can be reprogrammed by inserting genes for particular transcription factors ✓, which either prevent or encourage the transcription of specific sets of genes.

ⓔ 2/2 marks awarded This is an adequate answer for a 2-mark question. The student begins by clearly stating what reprogramming is, and then explains how this can be done, using correct terminology.

> **(b) (iii)** It would be good because it is difficult to get stem cells from embryos as there aren't many available embryos ✓ and they might have to be killed ✓, which many people disapprove of. There are lots of testes and it wouldn't do any real harm ✓ to a man to have some cells taken from his testes.

ⓔ 3/5 marks awarded A bit more needs to be made of the ethical objections to using embryos, in order to get a mark for that. Other points that could have been made include the fact that if the person to be treated was male, then his own testes could provide the stem cells. When the stem cells were placed into his body, it would accept them because they would be his own cells. If cells from an embryo were used, his immune system would probably reject them unless he was treated with immunosuppressant drugs.

Question 7

Clostridium difficile is a bacterium that lives naturally in the alimentary canal of many people. However, in people who are weakened by illness, or who have been taking antibiotics that have affected their normal gut bacterial populations, *C. difficile* populations in the gut may increase. The bacterium produces toxins that may cause serious diarrhoea and inflammation of the colon. In severe cases, death may result.

(a) The graph shows the number of reported cases of infection with *C. difficile* in hospitals in England between 1990 and 2005.

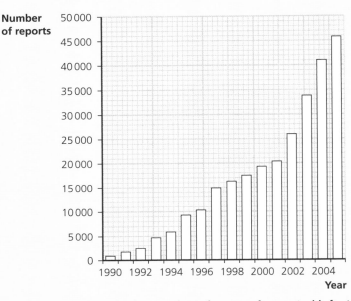

    (i) Describe the changes in the number of cases of reported infection with *C. difficile* between 1990 and 2005.   (2 marks)

    (ii) Suggest *two* reasons for the changes you have described.   (2 marks)

(b) Mild cases of *C. difficile* infection are usually treated with the antibiotic metronidazole, which inhibits bacterial DNA synthesis. More severe cases are treated with the antibiotic vancomycin, which inhibits the formation of cross-links in the bacterial cell wall. Both of these antibiotics are bactericidal if given in sufficiently large doses.

    (i) Suggest why neither metronidazole nor vancomycin have harmful effects on human cells.   (2 marks)

    (ii) Explain the meaning of the term bactericidal.   (1 mark)

    (iii) Vancomycin is usually given intravenously (into a blood vessel), but for *C. difficile* infections it is normally given by mouth. Suggest why this is done.   (1 mark)

(c) Since the data shown in the graph above were collected, there has been some reduction in the number of cases of *C. difficile* contracted by patients in hospitals. These are thought to be due to implementation of a code of practice for preventing and treating these infections.

Suggest how each of these guidelines in the code of practice may have helped to reduce the number of *C. difficile* infections in hospitals.

(i) All healthcare workers should wash their hands with soap and water before and after contact with patients with suspected *C. difficile* infection. (2 marks)

(ii) Antibiotics used for the treatment of *C. difficile* infection should not be broad-spectrum ones (that is, antibiotics that kill a wide range of bacteria), but should specifically target *C. difficile*, and antibiotics should not be used at all unless clearly necessary. (3 marks)

Total: 13 marks

🄔 There is a lot of reading in this question. Read *all* of it, very carefully, before you begin to construct your answers.

Student A

(a) (i) There has been a fairly steady rate of increase ✓ since 1990, from about 250 to just over 45 000 a year. This is 180 times ✓ as many cases.

🄔 **2/2 marks awarded** Two correct statements, including some manipulation of the data. 'Fairly steady' is just worth a mark, although a careful consideration of the shape of the graph actually shows that the rate of increase is greater in more recent years.

(a) (ii) People are living longer, so there are more old people around and they are the ones most likely to get infected ✓. And people don't wash their hands properly in hospitals.

🄔 **1/2 marks awarded** The first suggestion is valid but the second one (although it may well be an important factor in spreading the bacterium) cannot explain the change in incidence, unless people wash their hands less now than they used to.

(b) (i) Because our cells are different from bacteria.

🄔 **0/2 marks awarded** There is not enough detail here for any marks to be awarded.

(b) (ii) Kills bacteria ✓.

🄔 **1/1 mark awarded**

(b) (iii) Because the bacterium infects the alimentary canal, so it is quicker for the antibiotic to reach it ✓.

🄔 **1/1 mark awarded** This is correct.

(c) (i) This would remove any bacteria from their hands ✓ so they won't give them to another patient when they touch them ✓.

ℯ 2/2 marks awarded There is just enough here for both marks.

(c) (ii) It is more likely that the antibiotics will kill *C. difficile*.

ℯ 0/3 marks awarded Student A is thinking along the right lines but the answer needs to contain more information.

Student B

(a) (i) The incidence has risen by about 42 000 cases over a 15-year period — a mean increase of 2800 cases a year ✓. The rise has been steeper since 2001 ✓.

ℯ 2/2 marks awarded These are two good points, one involving some manipulation of the figures.

(a) (ii) There may be more bacteria that have become resistant to antibiotics ✓, so there are more *C. difficile* around that have not been killed and this could increase the rate of infection. And other treatments in hospitals are better now than they used to be, so there are more ill people in hospitals who might have died before, and they have weak immune systems so they can't fight off the infection ✓.

ℯ 2/2 marks awarded These are two good suggestions, explained well.

(b) (i) The enzymes involved in the synthesis of DNA in bacteria are probably different ✓ from those in humans. And human cells don't have cell walls ✓.

ℯ 2/2 marks awarded Two correct points, specifically related to the information about the two antibiotics given in the question.

(b) (ii) It is something that kills bacteria ✓.

ℯ 1/1 mark awarded This is correct for a mark.

(b) (iii) The bacteria are in the alimentary canal, so the antibiotic will get to them quicker ✓.

ℯ 1/1 mark awarded Correct.

(c) (i) This removes bacteria from their hands ✓, especially if they use soap. So if they have picked up bacteria from one person, they will not transmit them ✓ to another one.

ℯ 2/2 marks awarded This is a good suggestion, well explained.

(c) (ii) *C. difficile* is more likely to grow if there aren't other 'friendly' bacteria in the digestive system, so if you use broad-spectrum antibiotics you might kill the friendly bacteria ✓ and actually make the infection worse ✓. By not using antibiotics at all you reduce the risk of resistant strains evolving ✓.

ⓔ 3/3 marks awarded Two good points, one of them well explained.

Question 8

A student was provided with a Petri dish containing nutrient agar, on which colonies of several different kinds of bacteria were growing. The student was asked to investigate the rate of growth of one species of these bacteria, using nutrient broth.

(a) Describe how the student should set up the culture of the chosen bacterial species in a conical flask of nutrient broth. (5 marks)

(b) The student decided to use a haemocytometer to count the number of bacteria in the broth every hour.
- He shook the conical flask containing the culture carefully, and then used a sterile pipette to take exactly $1\,mm^3$ of the bacterial suspension from the culture.
- He placed the $1\,mm^3$ sample into a sterile tube, and added $9\,mm^3$ of sterile distilled water, to make a 10^{-1} dilution. He then made further serial dilutions, so that the final dilution was 10^{-9}.
- He took a small volume from each dilution, and placed it onto a haemocytometer slide to count the bacteria.

(i) Did the student's methods allow him to make a total count or a viable count of the bacteria? Explain your answer. (1 mark)

(ii) Explain why the student made a series of dilutions of the sample of the culture that he took from the flask. (2 marks)

(c) For the 10^{-5} dilution made after 10 hours' incubation, the student counted 323 bacteria in 80 small squares of the haemocytometer. He knew that the volume of suspension covering one small square is $2.5 \times 10^{-4}\,mm^3$.
Calculate the number of bacteria in $1\,mm^3$ of the culture in the conical flask. Show your working. Express your answer in standard form. (4 marks)

Total: 12 marks

ⓔ This question tests your knowledge of a procedure that you should have carried out yourself. Certainly, if you have had hands-on experience of this practical technique, you will be at an advantage when answering the questions. The calculation in (c) is tricky, and it will be important to show each step in your working clearly.

Questions & Answers

> **Student A**
>
> **(a)** Sterilise an inoculating loop, then touch one of the bacterial colonies growing on the agar ✓. Dip the loop into some nutrient broth in a conical flask. Make sure that everything is sterile before you begin. Then shake the flask and put it into an incubator. The bacteria will take a long time to start growing, which is the lag phase. Then they will grow quickly, which is the log phase. Then they will run out of nutrients and their death rate will equal their birth rate, so the graph flattens out.

ⓔ **1/5 marks awarded** The description of the technique is basically correct, but there is not enough detail to gain more than 1 mark. For example, student A should make it clear that the nutrient broth and flask should be prepared before the sample is taken from the culture with the inoculating loop. The answer could also describe how the sample is taken from the culture with the loop. There is no mention of covering the opening of the flask. The impression given is that student A has not actually carried out this technique. The last three sentences of the answer are not relevant to the question, which does not ask about the pattern of growth expected in the culture.

> **(b) (i)** Total count, because it counts all of the bacteria in the flask.

ⓔ **0/1 marks awarded** This is correct as far as it goes but the answer does not properly explain that by 'all' the bacteria, this means both those that are living and those that are dead.

> **(b) (ii)** He wanted to make a lot of readings to increase the reliability of the results.

ⓔ **0/2 marks awarded** This is not correct. Dilutions are made so that you can choose the most appropriate one for counting the bacteria on the haemocytometer grid; you don't know which will be the best dilution until you have looked at the sample on the haemocytometer under the microscope.

> **(c)** 323 bacteria in 80 squares so there are 4.03 bacteria in 1 square.
>
> So there are 4.03 bacteria in 2.5×10^{-4} mm^3.
>
> So in the flask there must be $4.03 \times 2.5 \times 10^{-4} \times 10^5 = 10.075 \times 10^1$.

ⓔ **0/4 marks awarded** Student A has struggled with this calculation. There is no need to calculate the number of bacteria in 1 square; it is better to calculate the volume over 80 squares (which works out as 0.02 mm^3), in which we know there are 323 bacteria. The student then needs to use this value to find the number of bacteria in 1 mm^3 of the diluted sample, and then from there they can work out the number in 1 mm^3 of the undiluted sample.

Student B

(a) First, make up some sterile nutrient broth and put it into a sterile conical flask ✓. Put some sterile cotton wool in the top and let it cool down ✓. Then take an inoculating loop and hold it at the top of the blue cone in a Bunsen flame until it is red hot ✓. Lift the lid of the Petri dish just a little and gently touch the chosen colony of bacteria with one edge of the loop ✓ (which will have cooled down enough by then not to kill the bacteria). Put the loop into the broth in the flask and shake it around a bit ✓. You can do this several times using the same colony if you don't think you put enough bacteria in the first time. Then put the flask into an incubator at the temperature you want to culture the bacteria ✓. You shouldn't do this at 37°C because you don't want to culture bacteria that are adapted to grow inside humans.

ⓔ 5/5 marks awarded This answer suggests that Student B has done this procedure before, because there is a lot of good practical detail in it, and it follows a clear sequence.

(b) (i) It was a total count, because using the haemocytometer you cannot tell whether the cells you are counting are dead or alive ✓.

ⓔ 1/1 marks awarded Correct. There is not usually a mark for making a choice between two alternatives; here the mark is for explaining the reason for the choice.

(b) (ii) He didn't know how many bacteria there would be in $1\,mm^3$ of the sample. When you put them onto the haemocytometer, you can only count the bacteria if there aren't too many of them ✓. Probably, if he put the undiluted sample on, the bacteria would have been all piled up on top of one another on the haemocytometer and couldn't be counted.

ⓔ 1/2 marks awarded This is correct, and explains why the student diluted the sample, but it does not explain why he made a *series* of dilutions. Student A could also have explained that you don't know which dilution will give you a suitable number of bacteria to count on the haemocytometer grid, so you make a whole range of them and then decide which one to count.

(c) volume of suspension over 80 squares = $80 \times 2.5 \times 10^{-4}\,mm^3$

$= 2 \times 10^2\,mm^3$ or $0.02\,mm^3$ ✓

So there were 323 bacteria in $0.02\,mm^3$ of the sample.

So there would be $323/0.02 = 16\,150$ bacteria in $1\,mm^3$ ✓.

This had been diluted by 10^{-5} so the actual number of bacteria in $1\,mm^3$ of the undiluted sample would be $16\,150 \times 10^5$ ✓.

$= 1.615 \times 10^9$ bacteria mm^{-3} ✓

ⓔ 4/4 marks awarded The calculation is set out clearly and is entirely correct.

Question 9

Soya beans are legumes that are grown for human consumption and animal fodder. They contain relatively high amounts of protein and lipid, as well as dietary fibre and a range of vitamins and minerals. They grow well even in poor soils, as they have a symbiotic relationship with nitrogen-fixing bacteria, which convert nitrogen from the air to nitrates.

(a) Explain how the relationship with nitrogen-fixing bacteria enables soya beans to grow well even in poor soils. (2 marks)

(b) Many different types of genetically modified soya beans have been produced. The most common method of transferring DNA into the soya bean cells involves the use of *Agrobacterium tumefaciens*.
Explain how *A. tumefaciens* can transfer DNA into soya bean cells. (5 marks)

(c) In order to investigate gene function in soya bean plants, 'knockout' plants have been produced.
Explain what is meant by a 'knockout' plant, and suggest how the production of such plants provides information about gene function. (2 marks)

Total: 9 marks

ⓔ The first part of this question asks you to recall work done much earlier in your course. Part (b) is relatively straightforward as long as you have learnt this topic, while part (c) requires you to recall knowledge that you should have about 'knockout' mice, and then use this in the context of soya beans.

Student A

(a) Nitrates are used for making proteins ✓ for growth, so this will help the plants grow better.

ⓔ **1/2 marks awarded** This is correct as far as it goes. For a second mark, student A would need to provide more information about how proteins help a plant to grow, or information about why the production of nitrates by the symbiotic bacteria is particularly useful in a poor soil.

(b) There are plasmids in *A. tumefaciens* and these ✓ can have other DNA put into them ✓ that the bacterium then puts into the plant cells. You mix the bacterium with the plant cells and the bacterium injects the plasmid into the cells ✓. Not all the cells will get the plasmid so you have to check which ones have got the new DNA and which ones haven't.

ⓔ **3/5 marks awarded** This answer gives a basic description, and contains just enough information to get 3 marks. More detail would make a great difference — for example, how is the 'other DNA' inserted into the plasmids? How can you check which plants cells have taken up recombinant plasmids?

(c) It is like a knockout mouse where there is a gene that doesn't work ✓, so you can tell what the gene does.

e **1/2 marks awarded** The statement that 'there is a gene that doesn't work' is just enough for 1 mark, but the answer does not tell us how this enables information to be gained about the function of that gene.

Student B

(a) The plants need nitrates to make proteins ✓, for making new cells. If the soil is poor, then it may not contain enough nitrates ✓ for the plants to grow well.

e **2/2 marks awarded** This is a good concise answer that makes two clear and relevant points.

(b) *A. tumefaciens* is a bacterium that infects plants and makes them grow tumours ✓. The bacteria contain small circular pieces of DNA, called plasmids ✓, that they insert into the plant cells. Genetic engineers can get the DNA that they want to insert into the plant, and put it into a plasmid. They do this by cutting the DNA and the plasmid with the same restriction enzyme ✓, so that the sticky ends of the DNA can make hydrogen bonds with the sticky ends of the plasmid ✓. Then DNA ligase ✓ is used to join the strands of DNA together. The bacteria are mixed with the plasmids so they take them up ✓, then the bacteria introduce the plasmids to the plant cells. The bacterium is being used as a vector ✓.

e **5/5 marks awarded** Student B tells us what *A. tumefaciens* is, and gives considerable relevant detail about the process by which it is used as a vector to introduce DNA to plant cells. There is excellent use of terminology. There are many other points that could have been made, but there is more than enough here for the maximum number of available marks — in fact, Student B could have written an answer half this length and still gained the same number of marks.

(c) A knockout plant is one in which a gene has been inactivated ✓. You can then compare this plant with one in which the gene is still working, and that helps you work out what the gene normally does ✓.

e **2/2 marks awarded** It would have been good to know how the gene would be inactivated, but nevertheless this is just enough for both marks.

Knowledge check answers

Knowledge check answers

1 Similarities: it contains the pentose sugar ribose; it contains a base (adenine); it contains phosphate. Differences: it contains three phosphate groups rather than one; it always contains the same base, whereas an RNA nucleotide can contain any of four bases (adenine, guanine, cytosine or uracil).

2 The addition of phosphate raises the energy level of the glucose molecule, allowing it to be split.

3 The carbon comes from glucose.

4 It is 'oxidative' because at the end of the electron transport chain the hydrogen ions and electron are oxidised to form water. It is 'phosphorylation' because ADP is phosphorylated (has a phosphate group added to it) to form ATP.

5 If no soda lime was included, then no carbon dioxide would be absorbed. During aerobic respiration of glucose, the volume of oxygen taken in equals the volume of carbon dioxide given out, so the manometer fluid would not move in either tube.

6 During anaerobic respiration, the organisms would give out carbon dioxide but not take in oxygen. As the carbon dioxide is absorbed by the soda lime, the manometer fluid would not move.

7 The absorption spectrum shows the wavelengths of light that can be absorbed. The action spectrum shows which wavelengths of light can be used. We would therefore expect the two to be very similar to one another.

8 The apparatus could be placed inside a thermostatically controlled water bath. In practice, this is tricky, because it would need to stand within a dry compartment with water around it.

9 $Rf = 5.9/9.8 = 0.60$. The pigment is chlorophyll a.

10 Both have two membranes surrounding them — an envelope. Both contain their own DNA and ribosomes. Both contain components of the electron transport chain on their membranes. Both have a background material containing various enzymes — the stroma in a chloroplast, the matrix in a mitochondrion.
They differ in that chloroplast membranes are organised into stacks called grana, whereas the inner mitochondrial membrane is folded to form cristae. Chloroplast membranes contain chlorophyll and other pigments, which are not present in mitochondria. Chloroplasts often contain starch grains, which are not found in mitochondria.

11 The energy is transferred to ATP and reduced NADP.

12 The substrates are RuBP and carbon dioxide. The product is GP.

13 They are produced in the light-dependent stage.

14 This reduces the chance of microorganisms falling onto the surface of the agar from the air.

15 7

16 $P_t = 5752$ $P_0 = 694$ $t = 4$ hours
$5752 = 694e^4k$
$8.29 = e^4k$
$\ln 8.29 = 4k$
$2.12 = 4k$
$k = 0.53$ hour^{-1}

17 Viruses do not have cells, so do not have cell walls that penicillin could damage. They do not have ribosomes and do not make their own proteins, so tetracycline could not harm them.

18 The cause of a disease is the organism that invades the body and produces symptoms; it is otherwise known as a pathogen. A vector is an organism that transfers the causative organism — the pathogen — from one host to another.

19 The mitochondria are sites for aerobic respiration, where ATP is produced by the Krebs cycle and oxidative phosphorylation; ATP is needed to provide energy for protein synthesis. Rough endoplasmic reticulum is the site of protein synthesis, particularly of proteins for export from the cell, including antibodies. The Golgi bodies process and package the antibodies, ready for secretion from the cell.

20 A vaccination containing attenuated bacteria would be useless, as the person may already have been infected with live bacteria, and there is no time for them to develop their own immunity against it. They will now have passive, artificial immunity.

21 For gel A, all the bands in the child's profile match up with either the mother's or the potential father's, so this man could be the child's father. In gel B, the child has a band that is not present in either the mother's or the father's results, so some other person must be the child's father.

22 Synthesising amylase requires energy and materials (ATP and amino acids), and so uses up resources. It is advantageous for the seed to conserve these resources until they are needed.

23 Some of the genes in the cells undergo epigenetic modifications, which prevents some genes being expressed.

24 There are only two copies of the insulin gene (DNA) in each cell. There are many copies of the mRNA, because this is being used to produce insulin. Moreover, a high proportion of the mRNA in the cell will be for insulin, because making insulin is the main purpose of the cell.

25 No. Some of the plasmids will just join up their broken ends again, without incorporating the extra DNA.

Index

endemic disease control 32–33
fungi, viruses and protoctists 31–32
immune response 33–37
microbial techniques 25–28
practicals 29
microorganisms, culturing 25–26
mitochondrion 10, 11
mosquitoes and malaria 32
mRNA splicing 41–42
multipotent stem cells 43

N

NADP, photosynthesis 16, 21, 22, 23
NAD, respiration 9, 10, 11, 12–13
natural immunity 36
neutrophils 33

O

oxidative phosphorylation 11–12

P

passive immunity 36
paternity testing 38, 40
pathogenic agents
    bacteria 30
    fungi, viruses and protoctists 31–32
penicillin 30
phagocytes 33
photolysis 21
photophosphorylation 21
    cyclic and non-cyclic 22
photosynthesis 16, 20–24
    light-dependent stage 21–22
    light-independent stage 22–23
    limiting factors 23–24
    rate of, measuring 18–19
photosystems 20, 22
phytochrome-interacting factor (PIF) 40–41
pigments, photosynthetic 16–17
    core practical 19–20
plasma cells 34
plasmids 31

in gene technology 45, 46–47
Plasmodium 32
pluripotent stem cells 43
polymerase chain reaction (PCR) 38, 39
promoter, DNA region 40, 41
protoctists 32
PSII photosystem 20, 21
PSI photosystem 20, 21, 22
pyruvate 9, 10, 12, 13, 14

R

recombinant cells, steps in producing 44–47
redox indicator 18
replica plating 46, 47
reprogrammed stem cells 44
respiration
    aerobic 8–12
    anaerobic 12–14
    measuring rate of 14–15
restriction endonucleases 44, 45
reverse transcriptase 45
RNA-associated silencing 42
RUBISCO 22, 23
RuBP (ribulose bisphosphate) 22, 23

S

scientific language 6
secondary immune response 36
soya beans, genetically modified 47
stem cells 43–44
stem rust 31
sterile agar plate 25, 26
sterile media 25
streak plate, making 26
stroma 20, 21, 22

T

T cells 33, 35–36
tetracycline 30, 46, 47
thylakoids 20, 21
time management in exams 7
totipotent stem cells 43
toxins 30

transcription factors 40–41
turbimetry 27

V

vaccination programmes 36–37
vector
    for a disease 32
    in gene technology 45, 46
viruses 31–32
    immune response against 35–36
    use as vectors in gene technology

W

wavelengths of light 16–17
    core practical 19

Z

Z-scheme 21–22